Cell Biology of the Ovary

Hidetaka Katabuchi · Takashi Ohba
Takeshi Motohara
Editors

Cell Biology of the Ovary

Stem Cells, Development, Cancer, and Clinical Aspects

Springer

Editors
Hidetaka Katabuchi
Department of Obstetrics and Gynecology
Faculty of Life Sciences
Kumamoto University
Kumamoto
Japan

Takashi Ohba
Department of Obstetrics and Gynecology
Faculty of Life Sciences
Kumamoto University
Kumamoto
Japan

Takeshi Motohara
Department of Obstetrics and Gynecology
Faculty of Life Sciences
Kumamoto University
Kumamoto
Japan

ISBN 978-981-10-7940-5 ISBN 978-981-10-7941-2 (eBook)
https://doi.org/10.1007/978-981-10-7941-2

Library of Congress Control Number: 2018940504

Printed on acid-free paper

This Springer imprint is published by the registered company Springer Nature Singapore Pte Ltd. part of Springer Nature
The registered company address is: 152 Beach Road, #21-01/04 Gateway East, Singapore 189721, Singapore

Preface

The ovaries are not only a part of the reproductive system, which contains ova, but also a part of endocrine system, which secretes sex steroid hormones such as estrogen and progesterone. In 1651, an English physician William Harvey postulated "ex ovo omnia—all that is alive comes from the egg" in his book *Exercitationes de Generatione Animalium*. His insightful words, written almost four centuries ago, express the importance of the information stored in the ova and its mechanism in the development of a living being.

This comprehensive book aims to provide extremely valuable reviews on the principles of the scientific biology of the ovary. Arranged thematically in sections, each chapter covers different but interrelated topics spanning across basic and clinical research. In this book, we present molecular mechanisms underlying ovarian development, especially focusing on the generation of primordial germ cells, gonadogenesis, and steroidogenesis in the gonads. Furthermore, we also provide the latest advances with respect to the functional role of reproductive epithelial stem cells and ovarian cancer stem cells. In addition, the most relevant and up-to-date clinical aspects of fertility preservation, particularly ovarian tissue cryopreservation in the context of oncofertility and management of epithelial ovarian cancer, including underlying molecular events, are described in detail.

The 32nd International Kumamoto Medical Bioscience Symposium entitled "The Ovary" was held on November 3, 2016, in Yamasaki Memorial Hall, Kumamoto University, and was one of the most notable symposia so far with great enthusiasm. This symposium provided us with motivation and impetus to create this book to consolidate comprehensive data from cutting-edge research covering a broad range of topics related to the ovary.

We are sincerely confident that this innovative book will prove to be exciting for readers from several disciplines, including young medical doctors, basic scientists, particularly those engaged in ovarian development and stem cell biology, as well as skilled obstetricians and gynecologists involved in various areas of academia, especially those specializing in reproductive endocrinology, oncofertility, and oncology. Finally, we would like to express our heartfelt gratitude and appreciation for all the leading expert authors.

Kumamoto, Japan

Hidetaka Katabuchi
Takashi Ohba
Takeshi Motohara

Contents

1 Reconstitution of Germ Cell Development In Vitro

Shihori Yokobayashi and Mitinori Saitou

Abstract

Mammalian germ cell development begins with the specification of primordial germ cells (PGCs) in the early phase of pregnancy. The development of PGCs to spermatozoa or mature oocytes involves complex pathways and takes a long time, e.g., more than 10 years in humans. Recent researches aimed at reconstituting germ cell development using pluripotent stem cells in vitro have achieved remarkable progress. Establishment of these in vitro systems is of great importance for addressing various biological questions on the nature of germ cells, as well as for studying the associated diseases and potentially for applied research.

Keywords

Germ cell development · Primordial germ cell-like cells · Pluripotent stem cells · In vitro induction

S. Yokobayashi (✉)
Department of Anatomy and Cell Biology, Graduate School of Medicine, Kyoto University, Kyoto, Japan

Japan Science and Technology Agency, Exploratory Research for Advanced Technology, Kyoto, Japan

Center for iPS Cell Research and Application, Kyoto University, Kyoto, Japan
e-mail: yokobayashi@anat2.med.kyoto-u.ac.jp

M. Saitou (✉)
Department of Anatomy and Cell Biology, Graduate School of Medicine, Kyoto University, Kyoto, Japan

Japan Science and Technology Agency, Exploratory Research for Advanced Technology, Kyoto, Japan

Center for iPS Cell Research and Application, Kyoto University, Kyoto, Japan

Institute for Integrated Cell-Material Sciences, Kyoto University, Kyoto, Japan
e-mail: saitou@anat2.med.kyoto-u.ac.jp

H. Katabuchi et al. (eds.), *Cell Biology of the Ovary*,
https://doi.org/10.1007/978-981-10-7941-2_1

1.1 Introduction

Germ cells are a unique type of cells that transmit genetic information through generations. In mammals, primordial germ cells (PGCs), the precursors of oocytes or spermatozoa, first emerge in the early postimplantation embryos (Fig. 1.1). The specification of PGCs entails repression of the somatic lineage program and reacquisition of potential pluripotency (reviewed in [12]). After specification, PGCs migrate toward the genital ridges through the hindgut epithelium, proliferate mitotically, and colonize in the developing gonads. During the course of development, PGCs reprogram their epigenetic states through genome-wide DNA demethylation and global changes of histone modifications, and erase parental-specific imprints and reactivate the inactive X chromosome in females. Subsequently, PGCs initiate sex-specific differentiation by arresting their cell cycle at G1/G0 phase in males or entering meiosis in females.

After the onset of sex differentiation, in the fetal testes, the arrested PGCs differentiate into prospermatogonia and acquire de novo DNA methylation, including

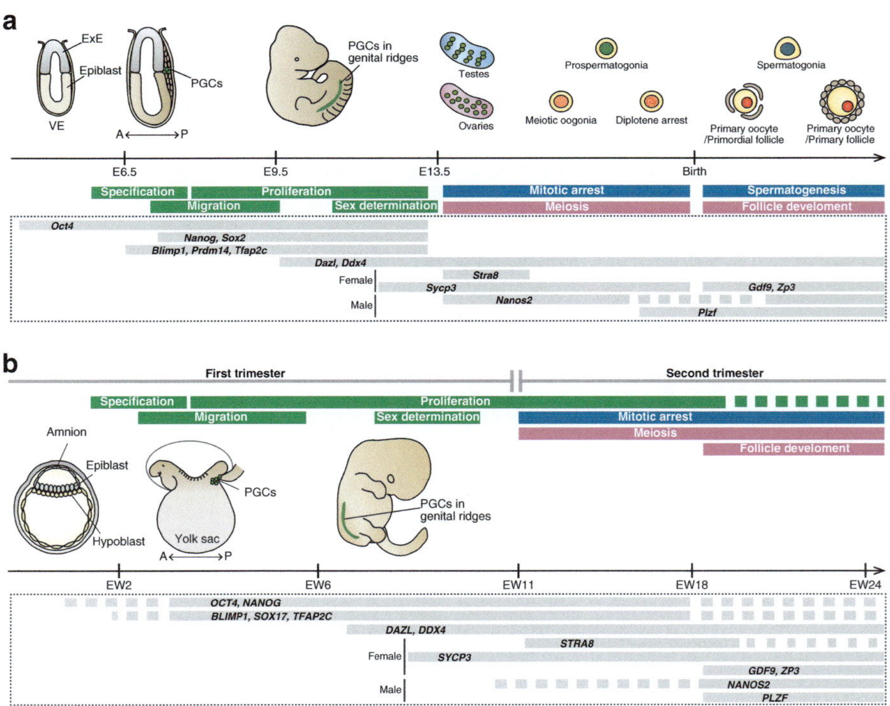

Fig. 1.1 (**a**) Schematic diagram of the mouse PGC development and subsequent spermatogenesis or oogenesis from embryonic day (E) ~6 to postnatal day (P) ~5 (top) and expression of key genes (bottom). *ExE* extraembryonic ectoderm, *VE* visceral endoderm, *A* anterior, *P* posterior. (**b**) Schematic diagram of the human PGC development and subsequent spermatogenesis or oogenesis from embryonic week (EW) ~2–24 (top) and expression of key genes (bottom) based on the following publications [1–11]

paternal imprint loci. These prospermatogonia then differentiate into spermatogonia to restart their cell cycle and repopulate in the basal compartment of seminiferous tubules after birth in mice or humans [13]. In female mice, the oogonia/oocytes proceed through meiotic prophase and are arrested at the diplotene stage, and at around birth, primordial follicles, complexes of individual oocytes, and a layer of surrounding granulosa cells are formed and start to develop. In association with follicle development, oocytes grow in size and acquire de novo DNA methylation at maternal imprint loci. In humans, oogenesis progresses rather asynchronously: PGCs, meiotic oogonia, and primordial follicles coexist in the fetal ovaries in the second trimester of pregnancy [14] (Fig. 1.1).

Over the last decades, significant progress has been made in the in vitro generation of germ cells using pluripotent stem cells (i.e., embryonic stem cells [ESCs] [15, 16] and induced pluripotent stem cells [iPSCs] [17, 18]) as starting materials. Remarkably, mouse PGC-like cells (mPGCLCs), induced from mESCs and miPSCs through the induction of epiblast-like cells (EpiLCs), have been shown to develop to gametes that contribute to healthy offspring [19–24]. Moreover, by using similar methodologies, human PGCLCs can be induced from hESCs and hiPSCs [25, 26]. These studies, together with in vivo studies on primates and ungulates [1, 27–32], also indicate the species differences in PGC development.

In this chapter, we will discuss the recent advances of in vitro reconstitution of germ cell development in mice and humans, and discuss future challenges for further progress in these research fields.

1.2 In Vitro Reconstitution of Mouse Germ Cell Development

1.2.1 Signaling Pathways for the Specification of Mouse Primordial Germ Cells

In mice, PGCs are induced in the most proximal posterior part of the epiblast upon the onset of gastrulation at around embryonic day (E) 6.0 and form a cluster of 30–40 cells within the extraembryonic mesoderm at E7.5 [33–35] (Fig. 1.1). At this stage, PGCs start to express *Blimp1/Prdm1*, *Prdm14*, and *Tfap2c*, as well as *Nanos3*, *Dnd1*, and *stella/Dppa3/Pgc7* [35–39]. PGCs also express and/or upregulate pluripotent factors such as *Oct4/Pou5f1*, *Sox2*, and *Nanog* to reacquire the potential pluripotency. In addition, PGCs upregulate *Kit*, a receptor for Kitl (also known as stem cell factor [SCF]), from E7.5 onward [36]. The Kit/SCF pathway plays a crucial role in the proliferation, migration, survival, and maturation of germ cells in embryonic and postnatal gonads [40].

Bone morphogenetic protein (BMP) 4, which is secreted from the extraembryonic ectoderm (ExE) from E6.0 onward, is an essential signal for determining germ cell fate [34]. Moreover, Ohinata et al. [41] revealed the signaling principle for the specification of mouse germ cell fate: the activation of WNT signals in the posterior visceral endoderm (VE) and epiblast cells in response to the ExE-secreted BMP signal, and the secretion of inhibitory signals, such as *Cer1*, from the anterior visceral endoderm (AVE) [42, 43]. When epiblasts isolated from E5.5 to E6.25 embryos are cultured under a floating condition in the presence of BMP4, PGC-like cells that express *Blimp1* and *stella* are induced in the cultured epiblasts [41], and addition of SCF and leukemia inhibitory factor (LIF) to the culture medium allows the growth/survival of the induced PGC-like cells [41, 44, 45]. Importantly, mPGC-like cells induced from the epiblast contribute to spermatogenesis when transplanted into a testis of a neonatal *W/Wv* mouse that lacks endogenous germ cells due to the mutation of *Kit* [41].

1.2.2 Induction of PGC-Like Cells (PGCLCs) from Mouse Pluripotent Stem Cells

mESCs cultured under a 2i condition, a defined condition with inhibitors for the mitogen-activated protein kinase (MAPK) and glycogen synthase kinase 3 (GSK3) [46], are in the ground state of naïve pluripotency and developmentally comparable to the preimplantation epiblast in terms of gene expression and DNA methylation profiles [47, 48]. By stimulation with activin A and basic fibroblast growth factor (bFGF) for 2 days, the 2i mESCs (and miPSCs) are induced into epiblast-like cells (EpiLCs) that show properties similar to the postimplantation epiblast. The EpiLCs bear the competence to be induced into mPGCLCs when cultured as floating aggregates in a medium containing BMP4, SCF, and LIF [19] (Fig. 1.2). The mPGCLCs at day 6 in the aggregate culture exhibit a transcriptome highly similar to that of E9.5 mPGCs at a migration phase. The function of mPGCLCs was validated in transplantation experiments into a testis of a neonatal *W/Wv* mouse; the transplanted mPGCLCs were successfully differentiated into spermatozoa that contributed to fertile offspring through intracytoplasmic sperm injection (ICSI) [19].

Epiblast stem cells (EpiSCs) are another type of mouse pluripotent stem cells and are derived from the epiblast of postimplantation embryos at E5.5–6.5 in the presence of activin A and bFGF [49, 50]. EpiSCs are in the primed state of pluripotency and are unable to contribute to chimeras when injected into blastocysts under conventional condition [51]. In contrast to EpiLCs, EpiSCs are refractory to PGCLC induction [19]. In fact, EpiSCs exhibit a transcriptome profile similar to that of the epiblast at the gastrulation stage (~E7.5) [52], while EpiLCs show properties similar to the pregastrulation epiblast (~E5.75) [19]. This clarifies that the competence for germ cell fate resides in the epiblast in the pregastrulation stage [41].

Blimp1, Prdm14, and Tfap2c are essential for PGC development, since the mutation of any of these genes results in a failure of PGC development [36–39, 53]. Blimp1 is required for repression of the somatic mesodermal genes in PGCs [36]; Prdm14 plays key roles for the reacquisition of potential pluripotency and execution of epigenetic changes in PGCs [39, 54, 55]. Forced expression of the three TFs (*Blimp1*, *Prdm14*, and *Tfap2c*) in EpiLCs is sufficient for inducing PGCLCs, bypassing the necessity of BMP4 and other cytokines [21]. Of note, overexpression of TFs in ESCs is not sufficient for PGCLC induction, illustrating the necessity of

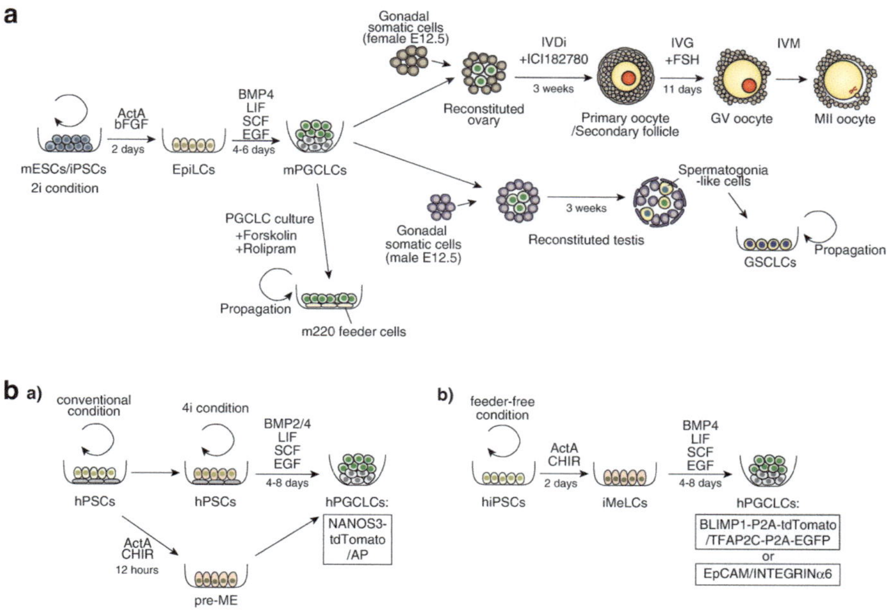

Fig. 1.2 (**A**) Schematic diagram of the in vitro system to induce mPGCLCs [19], to induce MII oocytes [22], to induce spermatogonia-like cells [23], and to propagate mPGCLCs [24]. mESCs/iPSCs under a condition with 2i (inhibitors for MAPK and GSK3) are induced into EpiLCs for 2 days by ActA (activin A) and bFGF. EpiLCs are subsequently induced into mPGCLCs in floating aggregates by BMP4 and other cytokines for 4–6 days. Female mPGCLCs are induced into primary oocytes by co-culture with somatic cells isolated from embryonic ovaries for 3 weeks (ICI182780, estrogen inhibitor; IVDi, in vitro differentiation). The primary oocytes are grown to GV (germinal vesicle) oocytes by FSH (follicle-stimulating hormone) (IVG, in vitro growth) and then matured into MII oocytes (IVM, in vitro maturation). Male mPGCLCs are induced into spermatogonia-like cells by co-culture with somatic cells isolated from embryonic testes for 3 weeks; and GSCLCs are subsequently derived from these spermatogonia-like cells. mPGCLCs are propagated when cultured on m220 feeder cells in the presence of forskolin and rolipram. (**B**) Schematic diagram of the in vitro system to induce hPGCLCs. (a) hPSCs (hESCs/iPSCs) cultured under a conventional condition are switched to a culture condition with 4i (inhibitors for MAPK, GSK3, p38, and JNK) and feeder cells, and then induced into hPGCLCs [25]. Alternatively, conventional hPSCs are induced into pre-ME (pre-mesodermal cells) by ActA and CHIR (CHIR99021) prior to induction of hPGCLCs [32]. AP denotes alkaline phosphatase activity. (b) hiPSCs under a feeder-free condition are induced into iMeLCs by ActA and CHIR for 2 days and then induced into hPGCLCs [26]

the EpiLC state in order to gain competence for the germ cell fate. The importance of the three TFs for the regulation of PGC genes has also been suggested in experiments using embryonal carcinoma cells [56]. T, a classical mesodermal factor induced by WNT signaling, directly regulates *Blimp1* and *Prdm14* in the presence of BMP signaling to facilitate their robust activation [57]. In addition, the pluripotent factor Nanog plays a role in upregulating *Blimp1* and *Prdm14* expression [58], suggesting that a positive feedback loop is active in the specified PGCs.

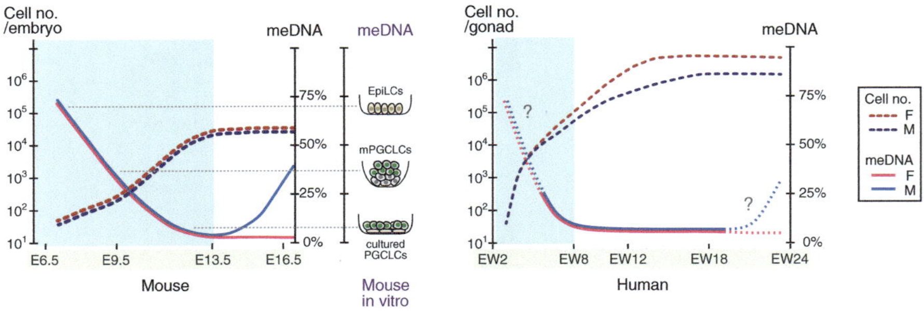

Fig. 1.3 Schematic diagram of the number of germ cells and the level of DNA methylation (mCpG) during the indicated fetal developmental period in mice (left) or humans (right). The level of DNA methylation in EpiLCs, mPGCLCs, and cultured mPGCLCs is presented (middle). *F* female, *M* male. The diagram is drawn based on the following publications [12, 14, 24, 29–31, 59–64]

After specification, PGCs undergo epigenetic reprogramming through genome-wide DNA demethylation and global changes of histone modifications [55]. Approximately 70% of CpG dinucleotides in the E6.5 epiblast contain 5-methylcytosine (5mC), and in the PGCs, the CpG methylation declines and reaches the lowest level at E13.5 in both sexes (~3%) [59–61] (Fig. 1.3). During the PGCLC induction, CpG demethylation proceeds from EpiLCs (~70%) to day 6 PGCLCs (~37%), reaching a 5mC level similar to that in E9.5 PGCs [62, 65], demonstrating that the process of epigenetic reprogramming is recapitulated in the mPGCLC induction system (Fig. 1.3). The level of methylated CpGs remains high in day 6 PGCLCs in the absence of Prdm14 (~68%), indicating the critical role of Prdm14 in the execution of DNA demethylation in PGCs [62]. Recently, Prdm14 was shown to form a complex with Cbfa2t2/Mtgr1, which has been reported as a histone deacetylase (HDAC)-recruiting corepressor [66–68]. Regarding histone modifications, a global increase of H3K27me3 and global decrease of H3K9me2 have been observed in the course of mPGCLC induction [69], which is reminiscent of in vivo PGCs in the migratory phase [55, 70]. Prdm14-Cbfa2t2 complex may facilitate the remodeling of chromatin states when cells are committed to the germ cell fate.

1.2.3 Completion of Mouse Oogenesis In Vitro

PGCs colonizing in genital ridges at around E10.0 undergo proliferation and start to express *Dazl* and *Ddx4/Vasa*, markers for gonadal PGCs and later germ cells (Fig. 1.1). In the in vitro culture system, day 6 mPGCLCs start to express *Dazl* and *Ddx4*, but only at a low level, and do not develop further under prolonged culture [19].

When mPGCLCs induced from female mESCs are co-cultured with gonadal somatic cells of E12.5 embryonic ovaries, mPGCLCs exhibit robust proliferation and developmental progression in the reconstituted ovaries [20]. In the reconstituted ovary cultures, mPGCLCs bear a transcriptome profile similar to that of E12.5 PGCs and demethylated DNA at imprint loci in 4–6 days, and further proceed into meiotic prophase in the prolonged culture. When the reconstituted ovaries are transplanted under the ovarian bursa of nude mice, mPGCLCs mature into fully grown oocytes that contribute to fertile offspring through in vitro maturation (IVM) and in vitro fertilization (IVF).

Morohaku et al. [71] have developed an in vitro culture system to generate fertile oocytes from PGCs using E12.5 female gonads as starting materials. This system has been applied to the reconstituted ovaries of mPGCLCs to promote further developmental progression in vitro; in the presence of an estrogen inhibitor, which prevents formation of follicles containing multiple oocytes [71], secondary follicle-like structures including single mPGCLC-derived oocytes are induced in 3 weeks [22] (Fig. 1.2). Subsequently, individual secondary follicles are manually separated and cultured in the presence of follicle-stimulating hormone (FSH) and other cytokines, where primary oocytes develop into fully grown germinal vesicle (GV) oocytes. When transferred to an IVM culture, the GV oocytes go through the first meiotic division, and the obtained MII oocytes are comparable to in vivo MII oocytes with respect to their diameters, and have established de novo DNA methylation at maternal imprint loci. The mPGCLC-derived MII oocytes give rise to two-cell embryos upon IVF, and the two-cell embryos, when transferred to surrogate mothers, give rise to viable pups at a rate of 3.5% or 0.9% when ESCs or iPSCs were used as starting materials, respectively [22]. This study was the first to report the successful reconstitution of an entire process of oogenesis in vitro, from mPSCs.

1.2.4 Induction of Spermatogonia-Like Cells from PGCLCs

In order to promote PGCLC differentiation into the male pathway in vitro, mPGCLCs were co-cultured with gonadal somatic cells of E12.5 embryonic testes. In the reconstituted testis cultures, mPGCLCs began to express Ddx4 in 2 weeks, and then became positive for Plzf/Zbtb16, a marker for prospermatogonia and spermatogonia, in 3 weeks [23] (Fig. 1.2). In vivo, a subpopulation of spermatogonia contributes to the generation of spermatogonial stem cells (SSCs) [72], and the prospermatogonia, spermatogonia, or SSCs can be propagated in vitro as a primary cell line, germline stem cells (GSCs), which are characterized by their capacity for colonizing and undergoing spermatogenesis in adult testes [73]. The induced spermatogonia-like cells in the reconstituted testis cultures were indeed competent for propagating as GSC-like cells; and these GSCLCs successfully colonized adult testes after transplantation, and a few of them (3 out of 15 lines) underwent spermatogenesis and contributed to fertile offspring [23] (Fig. 1.2). Given that the kinetics of differentiation of PGCLCs is protracted compared to that of PGCs in vivo [23], the method for reconstituting testes should be improved in the future. Collectively, this study presents a strategy for the derivation of SSC activity from mPSCs in vitro.

1.2.5 Propagation of PGCLCs In Vitro

mPGCLCs increase their numbers until day 6 or 8 of induction but wane thereafter in the floating aggregates [19, 24]. In Ohta et al. [24], chemical screening was performed using a system to culture mPGCLCs with m220 feeder cells, which express a membrane-bound form of SCF, known to support the survival of PGCs [44, 74], and two key factors, forskolin and rolipram, have been identified that support mPGCLC proliferation (Fig. 1.2). Forskolin and rolipram increase the intracellular level of cAMP by activating adenylyl cyclase (cAMP synthesis) or inhibiting Pde4 (cAMP hydrolysis), respectively. In fact, day 4 mPGCLCs cultured in the presence of both chemicals expand up to ~50-fold in 7 days [24]. Importantly, the cultured mPGCLCs maintain PGCLC properties during expansion and exhibit robust capacity for spermatogenesis through seminiferous tubule transplantation into neonatal testes [24], in stark contrast to embryonic germ cells (EGCs), which are derived from mPGCs in the presence of bFGF/LIF and exhibit mESC-like properties [75].

DNA demethylation proceeds in mPGCLCs during culture expansion; the methylation level at CpGs was reduced to ~6% after 7-day culture [24] (Fig. 1.3). This supports the idea that the replication/proliferation coupled passive mechanism majorly contributes to DNA demethylation in mPGCLCs, in the same manner as mPGCs [59, 60]. Moreover, demethylation-resistant regions that show slower kinetics of DNA demethylation have been identified in mPGCs [59], as well as in the cultured mPGCLCs, with a major overlap of those regions [24], indicating that the DNA demethylation kinetics unique to mPGCs are recapitulated in the culture system. Of note, demethylation-resistant regions are often associated with genes involved in meiosis and gamete generation [59, 62, 69], suggesting the importance of the acquisition of proper DNA demethylated states in PGCLCs. Therefore this mPGCLC culture system would be a useful tool for faithful reconstruction of germ cell development in vitro.

1.3 Induction of Human Germ Cell Fate In Vitro

1.3.1 In Vitro Systems to Induce Human PGCLCs from hPSCs

It was more than a hundred years ago that observations of human PGCs at premigratory phase were reported; in these studies, hPGCs have been observed in the endoderm of the wall of the yolk sac in 2.5-mm-long embryos (E23–26, Carnegie stage 11) and shown to have a distinct morphology, i.e., large size, spherical shape, and the presence of abundant glycogen granules in the cytoplasm [63, 76, 77] (Fig. 1.1). In subsequent studies, Politzer [78] and Witschi [79] described the distribution of PGCs in human embryos from the presomite stages at around embryonic week (EW) 3 (Carnegie stages 7–8) to EW5 (Carnegie stage 15) and observed their migration from the yolk sac endoderm to the primitive gonadal folds [12, 63]. Specification of hPGCs presumably occurs at around 2 weeks after fertilization, but the details remain veiled due to the difficulty in obtaining experimental materials.

Attempts to induce germ cells from human pluripotent stem cells (PSCs) in vitro have been performed for nearly two decades, and most of these studies have adopted undirected differentiation methods and focused on the expression of *DDX4* and/or *DAZL*, gonadal PGC markers (reviewed in [80]). In these studies, the induction of the marker-positive cells was often inefficient, and thus it was difficult to monitor the process of hPGC specification [12].

Conventional human/primate PSCs rely on FGF and activin/nodal signaling for self-renewal, exhibit flat-shaped morphology, and bear gene expression of a primed pluripotent state. Irie et al. [25] have utilized hPSCs that are maintained on feeder cells under a distinct culture condition (4i hPSCs) [81]. Based on the concept of mPGCLC induction [19], the 4i hESCs/hiPSCs are induced into hPGCLCs that express early PGC genes in a floating aggregate culture in the presence of BMP2/4 and other cytokines [25] (Fig. 1.2). Of note, hESCs maintained under a conventional culture condition (conventional hESCs) are scarcely induced into hPGCLCs [25]. On the other hand, Sasaki et al. [26] have utilized hiPSCs maintained under a feeder-free-defined condition with bFGF (feeder-free hiPSCs) [82]. The feeder-free hiPSCs exhibit the gene expression profile of a primed pluripotent state as in the conventional hiPSCs but in a more homogeneous manner [83]. The feeder-free hiPSCs are first differentiated into incipient mesoderm-like cells (iMeLCs) by stimulation of activin and WNT signals and subsequently differentiated into hPGCLCs in the floating aggregate culture in the presence of BMP4 and other cytokines [26] (Fig. 1.2).

The induced hPGCLCs express a substantial level of early PGC genes, such as *BLIMP1*, *TFAP2C*, *NANOS3*, *DND1*, and *KIT*, as well as *SOX17*, but express only a little amount of *STELLA* and do not express late PGC genes such as *DAZL* and *DDX4* [25, 26]. In addition, genes related to the naïve state of pluripotency, such as *KLF4* and *TFCP2L1*, are upregulated in hPGCLCs. *Sox2* is expressed throughout PGC development and PGCLC induction in mice, whereas the hPGCLCs repress *SOX2*, which is consistent with the previous reports showing the absence of *SOX2* in human gonadal PGCs and later germ cells [2, 84]. The hPGCLCs derived from either of the two methods described above exhibited a transcriptome profile that was globally similar to the gonadal PGCs of humans or cynomolgus monkeys (*Macaca fascicularis*, also known as crab-eating macaque), except for the gene classes that upregulate their expressions only later in migrating and/or gonadal PGCs, suggesting that the induced hPGCLCs share properties with human/primate PGCs of the early pre-migratory phase [25, 26]. The induction of hPGCLCs has also been reported by other groups [65, 85].

These in vitro systems allow us to address molecular pathways involved in hPGC specification. SOX17, a transcription factor required for the endoderm lineage, plays a critical role in specifying the human germ cell fate; none of the key PGC genes is upregulated in response to BMP signals in the absence of SOX17 [25]. This is in contrast to the dispensable role of mouse Sox17 for PGC specification [86]. Sox17 has been proposed to act as a substitute for Sox2 as a partner of Oct4 in the primitive endoderm lineage in mice [87], although it is unknown whether this is also the case with SOX17 in hPGCs. Expression of *SOX17* and downregulation of *SOX2* are also observed in cynomolgus PGCs as well as in porcine PGCs [28, 32]. In *BLIMP1*−/− hiPSCs, hPGCLC specification pathways are initiated with the formation of cells expressing *SOX17* and *TFAP2C* at day 2, but thereafter these cells decline in the aggregates [26]. The day 2 *BLIMP1*−/− PGCLCs exhibit the upregulation of genes related to gastrulation and neuronal differentiation, suggesting that BLIMP1 plays a conserved role in specifying germ cell fate through repression of the somatic differentiation program [26, 36–38]. On the other hand, downregulation of *SOX2* is not impaired in *BLIMP1*−/− PGCLCs [26], although this has been suggested in a previous study [88]. During the hPGCLC induction, BMP signaling likely plays a dominant role for the repression of *SOX2* [89]. The role of PRDM14 in human PGCs remains unclear; partial depletion of PRDM14 did not affect the induction of hPGCLCs [85]. In the course of hPGCLC induction, *PRDM14* is expressed at a rather low constant level [25, 26], which agrees with the observation in cynomolgus monkeys, where *PRDM14* was continuously expressed from the epiblast at around E7 until at least E36 in the developing PGCs [27, 28].

In vivo studies using cynomolgus monkeys have provided us with valuable insights into the germ cell development in primates. At E20, cynomolgus embryos are in the presomite stages (Carnegie stage 8) and exhibit primitive streaks, and cyPGCs, which are positive for SOX17, TFAP2C, and BLIMP1, are localized predominantly within/around the posterior parietal endoderm at the base of the allantois [28], consistent with the observations in early studies of human embryos. By using SOX17 and TFAP2C as markers, the nascent cyPGCs were detected at the earliest in E11 embryos at the amnion juxtaposed with the cytotrophoblast layer, and the location of the observed cyPGCs was subsequently shifted toward the posterior part in the amnion until E17 [28]. In primates including marmosets, rhesus macaques, and humans, the amnion is formed directly from the preimplantation epiblast before gastrulation, while in mice, amnion formation is initiated later at the onset of gastrulation (reviewed in [90]). At E11, *OCT4*, *SOX2*, and *NANOG* are uniformly expressed in the epiblast as well as the amnion, indicating that cyPGCs expressing *SOX17* and *TFAP2C* are induced from the amnion, which presumably bears gene expression properties similar to the epiblast [28]. The expressions of BMP4 and WNT3A preceding the emergence of the nascent cyPGCs are observed in the amnion at E11 embryos [28]. This may suggest that the signaling principle for PGC specification is highly conserved among species, although its spatial regulation is diverged due to species differences in embryonic morphology and lineage segregation programs in the early developmental stages.

1.3.2 Competence for the Induction of Human PGCLCs

Comparison of transcriptome profiles has shown that the feeder-free hiPSCs [82] are in the intermediate state between mEpiLCs and mEpiSCs in terms of the expressions of developmental genes [26], implying that the hiPSCs bear properties developmentally earlier than those of mEpiSCs [52]. More recently, the transcriptomes of human and monkey PSCs have been compared with pre- and postimplantation epiblasts of cynomolgus monkeys by single-cell RNA sequencing analyses; the profile of hiPSCs, as well as cyESCs, showed the highest similarity with late postimplantation epiblast cells before gastrulation [27].

In the iMeLCs induced from the feeder-free hiPSCs, pluripotency genes are expressed at the same level as in hiPSCs, but mesodermal genes such as *T*, *MIXL1*, *EOMES*, and *SP5* are upregulated [26]. The formation of floating aggregates directly from hiPSCs can lead to the induction of hPGCLCs but is accompanied by a massive cell loss [26], indicating that iMeLCs are in an optimal state for the induction of hPGCLCs. When the method in Sasaki et al. [26] was applied to many other hiPSC clones, the induction efficiency of hPGCLCs varied among the clones, and, interestingly, the clonal variations in the hPGCLC induction efficiency were reflected in the gene expression states of the iMeLCs [91]. In particular, the expression levels of *EOMES*, *MIXL1*, or *T* in the iMeLCs were positively correlated with the efficiency of subsequent hPGCLC induction [91]. On the other hand, the expression levels of these three genes were continuously increased when the iMeLC culture was excessively prolonged, and the resulting cells were no longer compatible for inducing hPGCLCs [26]. These observations suggest that the upregulation of these genes to desirable levels during iMeLC induction is likely a key to success in the subsequent hPGCLC induction. Of note, the 4i hESCs already exhibit the expression of *MIXL1* and *T* in an undifferentiated state in a heterogeneous manner [25], implying that 4i hESCs have already obtained characteristics similar to iMeLCs. Furthermore, a recent study reported the similar result that conventional hiPSCs gain competency for inducing PGCLCs through the induction into mesendoderm precursors (pre-ME) by stimulation with activin A and CHIR99032, and the pre-ME exhibit moderate upregulation of *EOMES*, *MIXL1*, and *T* [32] (Fig. 1.2).

T plays an important role in mPGC specification [57]. On the other hand, Eomes and Mixl1 are known for their essential roles in endoderm differentiation during the primitive streak formation in mice [92, 93] and are rapidly downregulated in the specified PGCs from E7.25 onward [36]. Similarly, the specified cyPGCs at E13 embryos exhibit substantial expression of *T* and low expression of *MIXL1* and *EOMES* [28]. In the systems to induce definitive endodermal cells from hESCs, knockdown of *EOMES* expression leads to a failure of differentiation [94]; moreover, ChIP-seq analysis has shown that EOMES associates with the enhancer regions of definitive endodermal genes, including *SOX17*, via the interaction of EOMES with SMAD2/3, downstream effectors of activin/TGFb signaling [89, 94, 95]. These studies suggest that EOMES plays a direct role in the transcriptional activation of *SOX17*. A recent study has demonstrated that EOMES is indeed required for the activation of *SOX17* in the course of hPGCLC differentiation, while deletion of T did not alter the induction of hPGCLCs: the transcriptional circuitry for PGCLC specification, in terms of key transcription factors and their hierarchies, diverges evolutionarily between mice and humans [96].

In most mammals, random X-chromosome inactivation (XCI) occurs in females to compensate the dosage of X-linked genes. In human blastocysts, both X chromosomes are transcriptionally active in the preimplantation epiblast [97, 98], and XCI is established during postimplantation development [99]. It has been reported that female hiPSCs retain XCI, which happens to be eroded, i.e., reactivated partially in a stochastic fashion, during culture [100–102]. On the other hand, hESCs exhibit line-to-line variations in X chromosome states upon derivation from blastocysts and also are subjected to erosion of XCI similar to hiPSCs [103–105]. Recently it has been demonstrated that human ESCs cultured under 5i/L/FA or t2iL + Gö conditions [106, 107] activate both X chromosomes [108], as in preimplantation epiblasts or mouse PSCs in naïve pluripotent states, but also exhibit genome-wide DNA hypomethylation including imprint loci, which is inconsistent with the situation in vivo [109, 110]. It remains to be elucidated whether/to what extent these hPSC variations in XCI states influence their differentiation potential and/or competence for the PGC fate.

1.4 Future Perspectives for Reconstituting Human Germ Cell Development In Vitro

In the current situation, the induced hPGCLCs reflect the properties of hPGCs in the early phase of their development. The next step would be to further differentiate hPGCLCs into later PGC-like cells that express gonadal PGC markers. In particular, *DAZL* encodes a germ cell-specific RNA-binding protein and has been proposed to be a licensing factor for PGC differentiation; *Dazl*-deficient PGCs of female C57BL/6 mice fail to enter meiosis, and those of male C57BL/6 mice fail to arrest their cell cycle [111, 112], suggesting the importance of *DAZL* expression in hPGCLCs. As shown by co-culture experiments of mPGCLCs with gonadal somatic cells [20, 22, 23], the presence of gonadal somatic cells, if compatible, would greatly support the differentiation of hPGCLCs. In xenotransplantation assays, human germ cells isolated from fetal as well as postnatal testes are able to survive in the seminiferous tubule basement membrane of the recipient mouse over the long term, but so far there is no evidence of their differentiating further [13, 113–115]. It remains unknown whether gonadal environments (somatic cells) of mouse or other mammalian species would be eligible to promote differentiation of human germ cells or hPGCLCs. Alternatively, one idea would be to induce gonadal somatic cell-like cells from hPSCs. Precursors of somatic cells in the gonad arise from some proliferative cells expressing *SF1/Nr5a1* in the coelomic epithelium (CE) at around E10.0 in mice [116, 117]. The CE originates from the intermediate mesodermal lineage, which contributes to the generation of much of the urogenital system, including the mesonephros and metanephros [118]. An improved understanding of early gonadogenesis would clearly be important for the generation of such an in vitro system.

One of the hallmark features of the germline is that it reprograms epigenetic states. During hPGCLC induction, a de novo DNA methyltransferase, *DNMT3B*, is downregulated from day 2 onward, and a decrease in the 5mC level is observed in day 8 hPGCLCs [26], suggesting that demethylation of DNA has been initiated and has partially progressed in the hPGCLCs. von Meyenn et al. reported that the level of CpG methylation in their hPGCLCs was ~68% at day 4, and it gradually decreased to ~55% at day 12 [65]. In hPGCs in vivo, the methylation level decreased to ~25% by EW5.5 and further to 8%, nearly the lowest level, after EW7 [29–31] (Fig. 1.3).

Given that mPGCs reach their lowest level of CpG methylation at E13.5 [59, 61], the kinetics of DNA demethylation is much slower in hPGCs, by fivefold if simply calculated (from E7.0 to E13.5 vs. from EW2–3 to EW7–8) [65] (Fig. 1.3). Regarding PGC proliferation, the number of hPGCs has been described as ~40,000 or ~20,000 per gonad at EW7 in females and males, respectively, and the doubling time was estimated as 5 days [64]. On the other hand, the number of mPGCs reaches ~25,000 at E13.5 [12]. These observations imply that the slower kinetics of DNA demethylation in hPGCs might be due to the slower rate of their proliferation. It would be important to create an in vitro condition that allows the induced hPGCLCs to proliferate, if necessary, at such a gradual rate, in order to recapitulate proper epigenetic changes.

Moreover, initiation of meiosis is a critical hallmark of the germline. In mice as well as in rats, differentiation from mitotic PGCs to meiotic oogonia in female gonads occurs synchronously, in an anterior to posterior wave [119, 120]. In humans, meiotic oogonia start to be observed from EW11–12 onward, while undifferentiated/mitotic oogonia expressing early PGC genes are still present at EW30, and thus more heterogeneous germ cell populations coexist in the fetal ovaries [1, 3, 14, 121] (Fig. 1.1). Interestingly, undifferentiated oogonia tend to be found at closer proximity to the ovarian surface epithelium than meiotic oogonia, suggesting the existence of a signaling wave from the medulla to cortical (dorsal to ventral) region in human fetal ovaries [3, 121]. Reconstitution of such signaling environments inducing meiosis would definitely be a key challenge for future progress in the in vitro development of human germ cells.

Conclusions

Due to the recent remarkable progress in the field, our understanding about the induction of the mammalian germ cell fate has greatly deepened, but there is still a long way to go with many challenges to be overcome before the complete reconstitution of gametogenesis in vitro. Given the implicit limitations as well as the great potential of research into the induction of human germ cell fate, it will be increasingly important to discuss the ethical aspects of all such efforts in the future.

Acknowledgments We would like to thank the members of our laboratory for their helpful discussions. This work was supported in part by a grant-in-aid from MEXT and by JST-ERATO.

References

1. Li L, Dong J, Yan L, Yong J, Liu X, Hu Y, et al. Single-cell RNA-seq analysis maps development of human germline cells and gonadal niche interactions. Cell Stem Cell. 2017;20:858–73. e4
2. Perrett RM, Turnpenny L, Eckert JJ, O'Shea M, Sonne SB, Cameron IT, et al. The early human germ cell lineage does not express SOX2 during in vivo development or upon in vitro culture. Biol Reprod. 2008;78:852–8.
3. Castrillon DH, Quade BJ, Wang TY, Quigley C, Crum CP. The human VASA gene is specifically expressed in the germ cell lineage. Proc Natl Acad Sci U S A. 2000;97:9585–90.

4. Stoop H, Honecker F, Cools M, de Krijger R, Bokemeyer C, Looijenga LH. Differentiation and development of human female germ cells during prenatal gonadogenesis: an immunohistochemical study. Hum Reprod. 2005;20:1466–76.
5. Anderson RA, Fulton N, Cowan G, Coutts S, Saunders PT. Conserved and divergent patterns of expression of DAZL, VASA and OCT4 in the germ cells of the human fetal ovary and testis. BMC Dev Biol. 2007;7:136.
6. Kerr CL, Hill CM, Blumenthal PD, Gearhart JD. Expression of pluripotent stem cell markers in the human fetal ovary. Hum Reprod. 2008;23:589–99.
7. Eckert D, Biermann K, Nettersheim D, Gillis AJ, Steger K, Jack HM, et al. Expression of BLIMP1/PRMT5 and concurrent histone H2A/H4 arginine 3 dimethylation in fetal germ cells, CIS/IGCNU and germ cell tumors. BMC Dev Biol. 2008;8:106.
8. Park TS, Galic Z, Conway AE, Lindgren A, van Handel BJ, Magnusson M, et al. Derivation of primordial germ cells from human embryonic and induced pluripotent stem cells is significantly improved by coculture with human fetal gonadal cells. Stem Cells. 2009;27:783–95.
9. Diedrichs F, Mlody B, Matz P, Fuchs H, Chavez L, Drews K, et al. Comparative molecular portraits of human unfertilized oocytes and primordial germ cells at 10 weeks of gestation. Int J Dev Biol. 2012;56:789–97.
10. Mamsen LS, Brochner CB, Byskov AG, Mollgard K. The migration and loss of human primordial germ stem cells from the hind gut epithelium towards the gonadal ridge. Int J Dev Biol. 2012;56:771–8.
11. Gkountela S, Li Z, Vincent JJ, Zhang KX, Chen A, Pellegrini M, et al. The ontogeny of cKIT+ human primordial germ cells proves to be a resource for human germ line reprogramming, imprint erasure and in vitro differentiation. Nat Cell Biol. 2013;15:113–22.
12. Saitou M, Miyauchi H. Gametogenesis from pluripotent stem cells. Cell Stem Cell. 2016;18:721–35.
13. Wu X, Schmidt JA, Avarbock MR, Tobias JW, Carlson CA, Kolon TF, et al. Prepubertal human spermatogonia and mouse gonocytes share conserved gene expression of germline stem cell regulatory molecules. Proc Natl Acad Sci U S A. 2009;106:21672–7.
14. Hartshorne GM, Lyrakou S, Hamoda H, Oloto E, Ghafari F. Oogenesis and cell death in human prenatal ovaries: what are the criteria for oocyte selection? Mol Hum Reprod. 2009;15:805–19.
15. Evans MJ, Kaufman MH. Establishment in culture of pluripotential cells from mouse embryos. Nature. 1981;292:154–6.
16. Thomson JA, Itskovitz-Eldor J, Shapiro SS, Waknitz MA, Swiergiel JJ, Marshall VS, et al. Embryonic stem cell lines derived from human blastocysts. Science. 1998;282:1145–7.
17. Takahashi K, Tanabe K, Ohnuki M, Narita M, Ichisaka T, Tomoda K, et al. Induction of pluripotent stem cells from adult human fibroblasts by defined factors. Cell. 2007;131:861–72.
18. Takahashi K, Yamanaka S. Induction of pluripotent stem cells from mouse embryonic and adult fibroblast cultures by defined factors. Cell. 2006;126:663–76.
19. Hayashi K, Ohta H, Kurimoto K, Aramaki S, Saitou M. Reconstitution of the mouse germ cell specification pathway in culture by pluripotent stem cells. Cell. 2011;146:519–32.
20. Hayashi K, Ogushi S, Kurimoto K, Shimamoto S, Ohta H, Saitou M. Offspring from oocytes derived from in vitro primordial germ cell-like cells in mice. Science. 2012;338:971–5.
21. Nakaki F, Hayashi K, Ohta H, Kurimoto K, Yabuta Y, Saitou M. Induction of mouse germ-cell fate by transcription factors in vitro. Nature. 2013;501:222–6.
22. Hikabe O, Hamazaki N, Nagamatsu G, Obata Y, Hirao Y, Hamada N, et al. Reconstitution in vitro of the entire cycle of the mouse female germ line. Nature. 2016;539:299–303.
23. Ishikura Y, Yabuta Y, Ohta H, Hayashi K, Nakamura T, Okamoto I, et al. In vitro derivation and propagation of spermatogonial stem cell activity from mouse pluripotent stem cells. Cell Rep. 2016;17:2789–804.
24. Ohta H, Kurimoto K, Okamoto I, Nakamura T, Yabuta Y, Miyauchi H, et al. In vitro expansion of mouse primordial germ cell-like cells recapitulates an epigenetic blank slate. EMBO J. 2017;36:1888–907.
25. Irie N, Weinberger L, Tang WW, Kobayashi T, Viukov S, Manor YS, et al. SOX17 is a critical specifier of human primordial germ cell fate. Cell. 2015;160:253–68.

26. Sasaki K, Yokobayashi S, Nakamura T, Okamoto I, Yabuta Y, Kurimoto K, et al. Robust in vitro induction of human germ cell fate from pluripotent stem cells. Cell Stem Cell. 2015;17:178–94.
27. Nakamura T, Okamoto I, Sasaki K, Yabuta Y, Iwatani C, Tsuchiya H, et al. A developmental coordinate of pluripotency among mice, monkeys and humans. Nature. 2016;537:57–62.
28. Sasaki K, Nakamura T, Okamoto I, Yabuta Y, Iwatani C, Tsuchiya H, et al. The germ cell fate of cynomolgus monkeys is specified in the nascent amnion. Dev Cell. 2016;39:169–85.
29. Gkountela S, Zhang KX, Shafiq TA, Liao WW, Hargan-Calvopina J, Chen PY, et al. DNA demethylation dynamics in the human prenatal germline. Cell. 2015;161:1425–36.
30. Guo F, Yan L, Guo H, Li L, Hu B, Zhao Y, et al. The transcriptome and DNA methylome landscapes of human primordial germ cells. Cell. 2015;161:1437–52.
31. Tang WW, Dietmann S, Irie N, Leitch HG, Floros VI, Bradshaw CR, et al. A unique gene regulatory network resets the human germline epigenome for development. Cell. 2015;161:1453–67.
32. Kobayashi T, Zhang H, Tang WWC, Irie N, Withey S, Klisch D, et al. Principles of early human development and germ cell program from conserved model systems. Nature. 2017;546:416–20.
33. Ginsburg M, Snow MH, McLaren A. Primordial germ cells in the mouse embryo during gastrulation. Development. 1990;110:521–8.
34. Lawson KA, Dunn NR, Roelen BA, Zeinstra LM, Davis AM, Wright CV, et al. Bmp4 is required for the generation of primordial germ cells in the mouse embryo. Genes Dev. 1999;13:424–36.
35. Saitou M, Barton SC, Surani MA. A molecular programme for the specification of germ cell fate in mice. Nature. 2002;418:293–300.
36. Kurimoto K, Yabuta Y, Ohinata Y, Shigeta M, Yamanaka K, Saitou M. Complex genome-wide transcription dynamics orchestrated by Blimp1 for the specification of the germ cell lineage in mice. Genes Dev. 2008;22:1617–35.
37. Ohinata Y, Payer B, O'Carroll D, Ancelin K, Ono Y, Sano M, et al. Blimp1 is a critical determinant of the germ cell lineage in mice. Nature. 2005;436:207–13.
38. Vincent SD, Dunn NR, Sciammas R, Shapiro-Shalef M, Davis MM, Calame K, et al. The zinc finger transcriptional repressor Blimp1/Prdm1 is dispensable for early axis formation but is required for specification of primordial germ cells in the mouse. Development. 2005;132:1315–25.
39. Yamaji M, Seki Y, Kurimoto K, Yabuta Y, Yuasa M, Shigeta M, et al. Critical function of Prdm14 for the establishment of the germ cell lineage in mice. Nat Genet. 2008;40:1016–22.
40. Mithraprabhu S, Loveland KL. Control of KIT signalling in male germ cells: what can we learn from other systems? Reproduction. 2009;138:743–57.
41. Ohinata Y, Ohta H, Shigeta M, Yamanaka K, Wakayama T, Saitou M. A signaling principle for the specification of the germ cell lineage in mice. Cell. 2009;137:571–84.
42. Rivera-Perez JA, Magnuson T. Primitive streak formation in mice is preceded by localized activation of Brachyury and Wnt3. Dev Biol. 2005;288:363–71.
43. Tam PP, Loebel DA. Gene function in mouse embryogenesis: get set for gastrulation. Nat Rev Genet. 2007;8:368–81.
44. Dolci S, Williams DE, Ernst MK, Resnick JL, Brannan CI, Lock LF, et al. Requirement for mast cell growth factor for primordial germ cell survival in culture. Nature. 1991;352:809–11.
45. Matsui Y, Toksoz D, Nishikawa S, Nishikawa S, Williams D, Zsebo K, et al. Effect of Steel factor and leukaemia inhibitory factor on murine primordial germ cells in culture. Nature. 1991;353:750–2.
46. Ying QL, Wray J, Nichols J, Batlle-Morera L, Doble B, Woodgett J, et al. The ground state of embryonic stem cell self-renewal. Nature. 2008;453:519–23.
47. Boroviak T, Loos R, Bertone P, Smith A, Nichols J. The ability of inner-cell-mass cells to self-renew as embryonic stem cells is acquired following epiblast specification. Nat Cell Biol. 2014;16:516–28.
48. Marks H, Kalkan T, Menafra R, Denissov S, Jones K, Hofemeister H, et al. The transcriptional and epigenomic foundations of ground state pluripotency. Cell. 2012;149:590–604.

49. Brons IG, Smithers LE, Trotter MW, Rugg-Gunn P, Sun B, Chuva de Sousa Lopes SM, et al. Derivation of pluripotent epiblast stem cells from mammalian embryos. Nature. 2007;448:191–5.
50. Tesar PJ, Chenoweth JG, Brook FA, Davies TJ, Evans EP, Mack DL, et al. New cell lines from mouse epiblast share defining features with human embryonic stem cells. Nature. 2007;448:196–9.
51. Nichols J, Smith A. Naive and primed pluripotent states. Cell Stem Cell. 2009;4:487–92.
52. Kojima Y, Kaufman-Francis K, Studdert JB, Steiner KA, Power MD, Loebel DA, et al. The transcriptional and functional properties of mouse epiblast stem cells resemble the anterior primitive streak. Cell Stem Cell. 2014;14:107–20.
53. Weber S, Eckert D, Nettersheim D, Gillis AJ, Schafer S, Kuckenberg P, et al. Critical function of AP-2 gamma/TCFAP2C in mouse embryonic germ cell maintenance. Biol Reprod. 2010;82:214–23.
54. Yabuta Y, Kurimoto K, Ohinata Y, Seki Y, Saitou M. Gene expression dynamics during germline specification in mice identified by quantitative single-cell gene expression profiling. Biol Reprod. 2006;75:705–16.
55. Seki Y, Hayashi K, Itoh K, Mizugaki M, Saitou M, Matsui Y. Extensive and orderly reprogramming of genome-wide chromatin modifications associated with specification and early development of germ cells in mice. Dev Biol. 2005;278:440–58.
56. Magnusdottir E, Dietmann S, Murakami K, Gunesdogan U, Tang F, Bao S, et al. A tripartite transcription factor network regulates primordial germ cell specification in mice. Nat Cell Biol. 2013;15:905–15.
57. Aramaki S, Hayashi K, Kurimoto K, Ohta H, Yabuta Y, Iwanari H, et al. A mesodermal factor, T, specifies mouse germ cell fate by directly activating germline determinants. Dev Cell. 2013;27:516–29.
58. Murakami K, Gunesdogan U, Zylicz JJ, Tang WWC, Sengupta R, Kobayashi T, et al. NANOG alone induces germ cells in primed epiblast in vitro by activation of enhancers. Nature. 2016;529:403–7.
59. Seisenberger S, Andrews S, Krueger F, Arand J, Walter J, Santos F, et al. The dynamics of genome-wide DNA methylation reprogramming in mouse primordial germ cells. Mol Cell. 2012;48:849–62.
60. Kagiwada S, Kurimoto K, Hirota T, Yamaji M, Saitou M. Replication-coupled passive DNA demethylation for the erasure of genome imprints in mice. EMBO J. 2013;32:340–53.
61. Kobayashi H, Sakurai T, Miura F, Imai M, Mochiduki K, Yanagisawa E, et al. High-resolution DNA methylome analysis of primordial germ cells identifies gender-specific reprogramming in mice. Genome Res. 2013;23:616–27.
62. Shirane K, Kurimoto K, Yabuta Y, Yamaji M, Satoh J, Ito S, et al. Global landscape and regulatory principles of DNA methylation reprogramming for germ cell specification by mouse pluripotent stem cells. Dev Cell. 2016;39:87–103.
63. De Felici M. Origin, migration, and proliferation of human primordial germ cells. Oogenesis. 2013;2013:19–37.
64. Mamsen LS, Lutterodt MC, Andersen EW, Byskov AG, Andersen CY. Germ cell numbers in human embryonic and fetal gonads during the first two trimesters of pregnancy: analysis of six published studies. Hum Reprod. 2011;26:2140–5.
65. von Meyenn F, Berrens RV, Andrews S, Santos F, Collier AJ, Krueger F, et al. Comparative principles of DNA methylation reprogramming during human and mouse in vitro primordial germ cell specification. Dev Cell. 2016;39:104–15.
66. Nady N, Gupta A, Ma Z, Swigut T, Koide A, Koide S, et al. ETO family protein Mtgr1 mediates Prdm14 functions in stem cell maintenance and primordial germ cell formation. elife. 2015;4:e10150.
67. Tu S, Narendra V, Yamaji M, Vidal SE, Rojas LA, Wang X, et al. Co-repressor CBFA2T2 regulates pluripotency and germline development. Nature. 2016;534:387–90.
68. Rossetti S, Hoogeveen AT, Sacchi N. The MTG proteins: chromatin repression players with a passion for networking. Genomics. 2004;84:1–9.

69. Kurimoto K, Yabuta Y, Hayashi K, Ohta H, Kiyonari H, Mitani T, et al. Quantitative dynamics of chromatin remodeling during germ cell specification from mouse embryonic stem cells. Cell Stem Cell. 2015;16:517–32.
70. Seki Y, Yamaji M, Yabuta Y, Sano M, Shigeta M, Matsui Y, et al. Cellular dynamics associated with the genome-wide epigenetic reprogramming in migrating primordial germ cells in mice. Development. 2007;134:2627–38.
71. Morohaku K, Tanimoto R, Sasaki K, Kawahara-Miki R, Kono T, Hayashi K, et al. Complete in vitro generation of fertile oocytes from mouse primordial germ cells. Proc Natl Acad Sci U S A. 2016;113:9021–6.
72. Kanatsu-Shinohara M, Shinohara T. Spermatogonial stem cell self-renewal and development. Annu Rev Cell Dev Biol. 2013;29:163–87.
73. Kanatsu-Shinohara M, Ogonuki N, Inoue K, Miki H, Ogura A, Toyokuni S, et al. Long-term proliferation in culture and germline transmission of mouse male germline stem cells. Biol Reprod. 2003;69:612–6.
74. Majumdar MK, Feng L, Medlock E, Toksoz D, Williams DA. Identification and mutation of primary and secondary proteolytic cleavage sites in murine stem cell factor cDNA yields biologically active, cell-associated protein. J Biol Chem. 1994;269:1237–42.
75. Matsui Y, Zsebo K, Hogan BL. Derivation of pluripotential embryonic stem cells from murine primordial germ cells in culture. Cell. 1992;70:841–7.
76. Fuss A. Uber extraregionare Geschlechtszellen bei einem menschlichen embryo von 4 Wochen. Anat Am. 1911;39:407–9.
77. Felix W. Die Entwicklung der Harn- und Geschlecht-sorgane. In: Keibel F, Mall FP, editors. Keibel- Mall Handbuch der Entwicklungsgeschichte des Menschen, vol. 2. Leipzig: Hirzel; 1911. p. 732–955.
78. Politzer G. Die Keimbahn des Menshen. Z Anat EntwGesch. 1933;100:331–6.
79. Witschi E. Migration of germ cells of human embryos from the yolk sac to the primitive gonadal folds. Contr Embryol Carnegie Inst. 1948;209:67–80.
80. Imamura M, Hikabe O, Lin ZY, Okano H. Generation of germ cells in vitro in the era of induced pluripotent stem cells. Mol Reprod Dev. 2014;81:2–19.
81. Gafni O, Weinberger L, Mansour AA, Manor YS, Chomsky E, Ben-Yosef D, et al. Derivation of novel human ground state naive pluripotent stem cells. Nature. 2013;504:282–6.
82. Nakagawa M, Taniguchi Y, Senda S, Takizawa N, Ichisaka T, Asano K, et al. A novel efficient feeder-free culture system for the derivation of human induced pluripotent stem cells. Sci Rep. 2014;4:3594.
83. Nakamura T, Yabuta Y, Okamoto I, Aramaki S, Yokobayashi S, Kurimoto K, et al. SC3-seq: a method for highly parallel and quantitative measurement of single-cell gene expression. Nucleic Acids Res. 2015;43:e60.
84. de Jong J, Stoop H, Gillis AJ, van Gurp RJ, van de Geijn GJ, Boer M, et al. Differential expression of SOX17 and SOX2 in germ cells and stem cells has biological and clinical implications. J Pathol. 2008;215:21–30.
85. Sugawa F, Arauzo-Bravo MJ, Yoon J, Kim KP, Aramaki S, Wu G, et al. Human primordial germ cell commitment in vitro associates with a unique PRDM14 expression profile. EMBO J. 2015;34:1009–24.
86. Hara K, Kanai-Azuma M, Uemura M, Shitara H, Taya C, Yonekawa H, et al. Evidence for crucial role of hindgut expansion in directing proper migration of primordial germ cells in mouse early embryogenesis. Dev Biol. 2009;330:427–39.
87. Aksoy I, Jauch R, Chen J, Dyla M, Divakar U, Bogu GK, et al. Oct4 switches partnering from Sox2 to Sox17 to reinterpret the enhancer code and specify endoderm. EMBO J. 2013;32:938–53.
88. Lin IY, Chiu FL, Yeang CH, Chen HF, Chuang CY, Yang SY, et al. Suppression of the SOX2 neural effector gene by PRDM1 promotes human germ cell fate in embryonic stem cells. Stem Cell Rep. 2014;2:189–204.
89. Loh KM, Ang LT, Zhang J, Kumar V, Ang J, Auyeong JQ, et al. Efficient endoderm induction from human pluripotent stem cells by logically directing signals controlling lineage bifurcations. Cell Stem Cell. 2014;14:237–52.

90. Boroviak T, Nichols J. Primate embryogenesis predicts the hallmarks of human naive pluripotency. Development. 2017;144:175–86.
91. Yokobayashi S, Okita K, Nakagawa M, Nakamura T, Yabuta Y, Yamamoto T, et al. Clonal variation of human induced pluripotent stem cells for induction into the germ cell fate. Biol Reprod. 2017;96:1154–66.
92. Hart AH, Hartley L, Sourris K, Stadler ES, Li R, Stanley EG, et al. Mixl1 is required for axial mesendoderm morphogenesis and patterning in the murine embryo. Development. 2002;129:3597–608.
93. Arnold SJ, Hofmann UK, Bikoff EK, Robertson EJ. Pivotal roles for eomesodermin during axis formation, epithelium-to-mesenchyme transition and endoderm specification in the mouse. Development. 2008;135:501–11.
94. Teo AK, Arnold SJ, Trotter MW, Brown S, Ang LT, Chng Z, et al. Pluripotency factors regulate definitive endoderm specification through eomesodermin. Genes Dev. 2011;25:238–50.
95. Kim SW, Yoon SJ, Chuong E, Oyolu C, Wills AE, Gupta R, et al. Chromatin and transcriptional signatures for nodal signaling during endoderm formation in hESCs. Dev Biol. 2011;357:492–504.
96. Kojima Y, Sasaki K, Yokobayashi S, Sakai Y, Nakamura T, Yabuta Y, Nakaki F, Nagaoka S, Woltjen K, Hotta A, Yamamoto T, Saitou M. Evolutionarily distinctive transcriptional and signaling programs drive human germ cell lineage specification from pluripotent stem cells. Cell Stem Cell. 2017;21(4):517–532.e5.
97. Okamoto I, Patrat C, Thepot D, Peynot N, Fauque P, Daniel N, et al. Eutherian mammals use diverse strategies to initiate X-chromosome inactivation during development. Nature. 2011;472:370–4.
98. Petropoulos S, Edsgard D, Reinius B, Deng Q, Panula SP, Codeluppi S, et al. Single-cell RNA-seq reveals lineage and X chromosome dynamics in human preimplantation embryos. Cell. 2016;165:1012–26.
99. Escamilla-Del-Arenal M, da Rocha ST, Heard E. Evolutionary diversity and developmental regulation of X-chromosome inactivation. Hum Genet. 2011;130:307–27.
100. Tchieu J, Kuoy E, Chin MH, Trinh H, Patterson M, Sherman SP, et al. Female human iPSCs retain an inactive X chromosome. Cell Stem Cell. 2010;7:329–42.
101. Tomoda K, Takahashi K, Leung K, Okada A, Narita M, Yamada NA, et al. Derivation conditions impact X-inactivation status in female human induced pluripotent stem cells. Cell Stem Cell. 2012;11:91–9.
102. Mekhoubad S, Bock C, de Boer AS, Kiskinis E, Meissner A, Eggan K. Erosion of dosage compensation impacts human iPSC disease modeling. Cell Stem Cell. 2012;10:595–609.
103. Silva SS, Rowntree RK, Mekhoubad S, Lee JT. X-chromosome inactivation and epigenetic fluidity in human embryonic stem cells. Proc Natl Acad Sci U S A. 2008;105:4820–5.
104. Vallot C, Ouimette JF, Makhlouf M, Feraud O, Pontis J, Come J, et al. Erosion of X chromosome inactivation in human pluripotent cells initiates with XACT coating and depends on a specific heterochromatin landscape. Cell Stem Cell. 2015;16:533–46.
105. Patel S, Bonora G, Sahakyan A, Kim R, Chronis C, Langerman J, et al. Human embryonic stem cells do not change their X inactivation status during differentiation. Cell Rep. 2017;18:54–67.
106. Theunissen TW, Powell BE, Wang H, Mitalipova M, Faddah DA, Reddy J, et al. Systematic identification of culture conditions for induction and maintenance of naive human pluripotency. Cell Stem Cell. 2014;15:471–87.
107. Takashima Y, Guo G, Loos R, Nichols J, Ficz G, Krueger F, et al. Resetting transcription factor control circuitry toward ground-state pluripotency in human. Cell. 2014;158:1254–69.
108. Sahakyan A, Kim R, Chronis C, Sabri S, Bonora G, Theunissen TW, et al. Human naive pluripotent stem cells model X chromosome dampening and X inactivation. Cell Stem Cell. 2017;20:87–101.
109. Pastor WA, Chen D, Liu W, Kim R, Sahakyan A, Lukianchikov A, et al. Naive human pluripotent cells feature a methylation landscape devoid of blastocyst or germline memory. Cell Stem Cell. 2016;18:323–9.

110. Theunissen TW, Friedli M, He Y, Planet E, O'Neil RC, Markoulaki S, et al. Molecular criteria for defining the naive human pluripotent state. Cell Stem Cell. 2016;19:502–15.
111. Lin Y, Gill ME, Koubova J, Page DC. Germ cell-intrinsic and -extrinsic factors govern meiotic initiation in mouse embryos. Science. 2008;322:1685–7.
112. Gill ME, Hu YC, Lin Y, Page DC. Licensing of gametogenesis, dependent on RNA binding protein DAZL, as a gateway to sexual differentiation of fetal germ cells. Proc Natl Acad Sci U S A. 2011;108:7443–8.
113. Nagano M, McCarrey JR, Brinster RL. Primate spermatogonial stem cells colonize mouse testes. Biol Reprod. 2001;64:1409–16.
114. Hermann BP, Sukhwani M, Hansel MC, Orwig KE. Spermatogonial stem cells in higher primates: are there differences from those in rodents? Reproduction. 2010;139:479–93.
115. Ramathal C, Durruthy-Durruthy J, Sukhwani M, Arakaki JE, Turek PJ, Orwig KE, et al. Fate of iPSCs derived from azoospermic and fertile men following xenotransplantation to murine seminiferous tubules. Cell Rep. 2014;7:1284–97.
116. Harikae K, Miura K, Kanai Y. Early gonadogenesis in mammals: significance of long and narrow gonadal structure. Dev Dyn. 2013;242:330–8.
117. Lin YT, Capel B. Cell fate commitment during mammalian sex determination. Curr Opin Genet Dev. 2015;32:144–52.
118. Taguchi A, Kaku Y, Ohmori T, Sharmin S, Ogawa M, Sasaki H, et al. Redefining the in vivo origin of metanephric nephron progenitors enables generation of complex kidney structures from pluripotent stem cells. Cell Stem Cell. 2014;14:53–67.
119. Hilscher B, Hilscher W, Bulthoff-Ohnolz B, Kramer U, Birke A, Pelzer H, et al. Kinetics of gametogenesis. I. Comparative histological and autoradiographic studies of oocytes and transitional prospermatogonia during oogenesis and prespermatogenesis. Cell Tissue Res. 1974;154:443–70.
120. Menke DB, Koubova J, Page DC. Sexual differentiation of germ cells in XX mouse gonads occurs in an anterior-to-posterior wave. Dev Biol. 2003;262:303–12.
121. Frydman N, Poulain M, Arkoun B, Duquenne C, Tourpin S, Messiaen S, et al. Human foetal ovary shares meiotic preventing factors with the developing testis. Hum Reprod. 2017;32:631–42.

The Role of Hedgehog-BMP4 Signaling in the Patterning of Coelomic Mesoderm and the Onset of Gonadogenesis

2

Takashi Yoshino

Abstract

All organogenesis is triggered by various cellular behaviors, based on early embryonic patterning. At the onset of the formation of the ovary or testis, a subset of coelomic epithelial cells undergo (1) epithelial-to-mesenchymal transition (EMT) and subsequent ingression, (2) gonadal differentiation, (3) maintenance of stem-like state at outer layer, and (4) acquisition of the capability to retain primordial germ cells (PGCs), in early embryos. The specific embryonic patterning responsible for these cellular behaviors represents a long-standing question in developmental biology. We addressed this question using chicken embryo, a suitable model system for cell-labeling experiments and spatiotemporally restricted gene manipulation. We found that the ventral part of medial lateral plate mesoderm (M-LPM), a precursor of coelomic epithelium, initiates gonadogenesis by undergoing ingression. By contrast, dorsal M-LPM did not undergo ingression or form gonad. Sonic Hedgehog (SHH)-BMP4 signaling establishes this dorsoventral pattern in M-LPM and initiates gonadogenesis. SHH protein is secreted from the endoderm, which is located ventral to LPM, and reaches ventral but not dorsal M-LPM. As a result, Hedgehog (Hh) signaling triggers several cellular behaviors involved in initiation of gonadogenesis in ventral M-LPM. Downstream of Hh signaling, BMP4 dynamically changes its expression and functions. It is well known that BMP4 is expressed throughout the whole LPM and establishes the mediolateral axis in the early embryo. Thereafter, at the onset of gonadogenesis, its expression is restricted to the ventral part and thus forms a dorsoventral pattern in the M-LPM in a Hh-dependent manner. As a result, BMP4 causes multiple cellular behaviors involved in gonad formation in the ventral M-LPM. Furthermore, other recent studies imply that these cellular

T. Yoshino
Department of Stem Cell Biology and Medicine, Faculty of Medical Science, Kyushu University, Fukuoka, Japan
e-mail: yoshino1@hgs.med.kyushu-u.ac.jp

H. Katabuchi et al. (eds.), *Cell Biology of the Ovary*,
https://doi.org/10.1007/978-981-10-7941-2_2

behaviors are regulated by several molecules, such as SNAIL2, NUMB, and SDF1, at the downstream of Hh-BMP4 signaling.

Keywords
Gonadal initiation · Coelomic epithelium · Sonic Hedgehog · BMP4 · Embryonic patterning · Lateral plate mesoderm · Chicken embryo

2.1 Introduction

The ovary, which plays several important roles in creating offspring, is constituted by a variety of cells. Granulosa cells support oocyte development, theca-interstitial cells secrete female hormone, and ovarian stem cells, which regenerate the ovary after ovulation, are located in the ovarian surface epithelium (OSE) [1–4]. In order to work appropriately, the ovary must form in the correct manner. The ovary develops from a sexually bipotential gonad by undergoing sex determination and sex differentiation. Extensive studies have revealed that Wnt4 and Foxl2 are key regulators of these processes [5–7].

By contrast, the fundamental characteristics of the ovary are established in the bipotential gonad, which appears as genital ridge at the both sides of the mesentery in the early embryo. The bipotential gonad consists of stem-like gonadal coelomic epithelium (G-CE), a precursor of OSE, and underlying mesenchyme [8, 9]. Primordial germ cells (PGCs), which differentiate to oocytes in the ovary, are maintained in this organ. Progenitors of bipotential gonad are thought to be located in the coelomic epithelium. To initiate gonadogenesis, these cells undergo epithelial-to-mesenchymal transition (EMT) and subsequent ingression, express several genes required for further gonadal development, keep stem-like state at outer layer, and acquire the capability to retain PGCs [5, 8–13]. A long-standing question in the field is how these processes are initiated and orchestrated.

Generally speaking, organogenesis occurs based on early embryonic axes (i.e., mediolateral axis) and embryonic patterning. However, it has remained unclear which embryonic axes regulate the trigger of gonadogenesis in coelomic epithelium or its precursor, lateral plate mesoderm (LPM). Recently, using chicken embryos, we uncovered the mechanisms of embryonic patterning underlying the onset and orchestration of gonadogenesis [14].

2.1.1 Contrasting Cellular Behaviors in the Dorsal and Ventral Parts of Medial LPM at the Onset of Gonadogenesis

In amniotes, the onset of gonadogenesis, which is caused by the ingression of coelomic epithelial cells, has been studied primarily in mouse embryos. These studies mainly focused on the fragmentation of basement membrane underlying coelomic epithelium as a sign of ingression [10, 11]. As a result, it has been thought that these cells initiate ingression after the embryo thickens (mouse day 10 embryo [E10] and chicken E3). On the other hand, we directly observed these cellular behaviors by taking advantage of the features of chicken embryos. Chicken embryos are a good model system for cell-labeling experiments [15, 16]. A chicken embryo can be observed by making a window at the top of an incubated egg. Therefore, target cells are easily labeled by fluorescent dyes, and embryonic development restarts after further incubation. We specifically labeled LPM cells (precursors of coelomic epithelial cells) of E2 embryo by injecting dye into the coelom and then observed how these cells initiated ingression and formed the bipotential gonad. Gonadal progenitor cells are expected to be induced at medial part of LPM (M-LPM) because several tissues (i.e., dorsal aorta and mesonephros), which surround the gonad in the later embryo, are located in the vicinity of this area (Fig. 2.1a) [17, 18]. Surprisingly, a cell-labeling experiment revealed that cells in the ventral M-LPM undergo ingression at E2, continue ingression, and express the gonadal marker GATA4 in E4.5 embryos (Fig. 2.1b, c). This result implies that gonadal progenitor cells appear at ventral M-LPM and initiate gonadogenesis in chicken E2 embryos (corresponding to mouse E8.5 embryos). By contrast, cells in the dorsal M-LPM neither undergo ingression nor express gonadal marker and instead become the mesonephric capsule, which covers the mesonephros [19, 20]. One of the important functions of the gonad is to attract and retain PGCs [21]. PGCs, which appear in gut mesenchyme by extravasation, are attracted to ventral M-LPM-derived gonadal cells but not to dorsal M-LPM-derived mesonephric capsular cells. Therefore, the machinery that establishes dorsoventral patterning in the M-LPM seems to initiate gonadogenesis (Fig. 2.1).

2.1.2 Hedgehog (Hh) Signaling Establishes Dorsoventral Pattern in M-LPM and Initiates Gonadogenesis

Furthermore, we demonstrated that the dorsoventral pattern in M-LPM is established by Hh signaling, which is activated by a secretory Hedgehog ligand such as SHH. Hh signaling regulates embryonic pattern formation, cell migration, and cell differentiation in multiple tissues [22, 23]. SHH is expressed in endoderm, which is formed ventral to M-LPM (Fig. 2.2a–c) [24]. It seems that SHH, secreted from endoderm, reaches and activates downstream signaling in the ventral part, but not the dorsal part, of M-LPM. *PATCHED*, a direct Hh signaling target gene, is specifically upregulated in the ventral part of M-LPM in E2 embryos (Fig. 2.2d–f).

This Hh signaling activation pattern plays central roles in the triggering and orchestration of gonadogenesis. It has been difficult to investigate the role of Hh signaling in the onset of gonadogenesis using knockout mice, largely because failure of Hh signaling in the whole body causes mouse embryos to die before the gonad appears [24]. We previously established a method for transfer genes into the M-LPM of chicken embryos in a spatiotemporally restricted manner by applying an in ovo electroporation technique [19]. By taking advantage of this method, we first investigated the role of Hh signaling in ventral M-LPM. We prevented Hh signaling

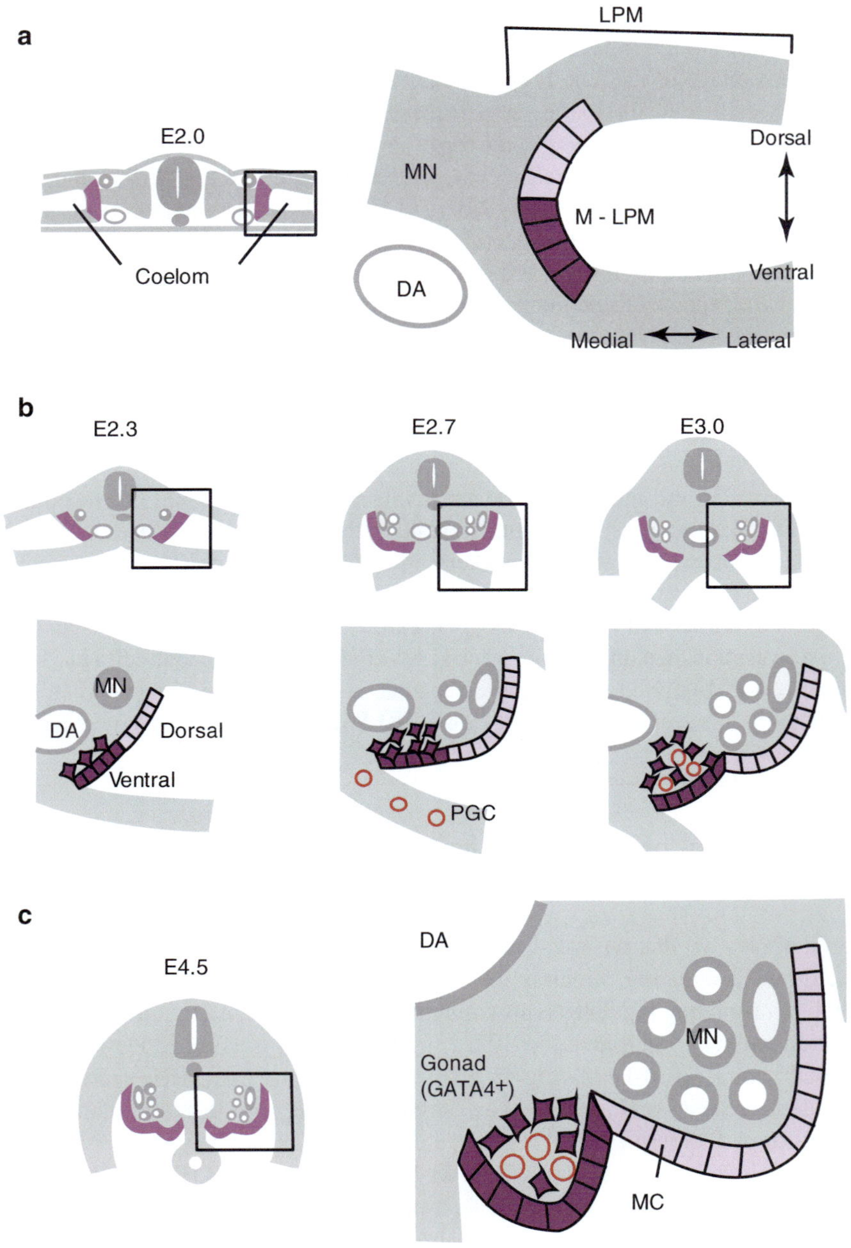

Fig. 2.1 Contrasting cellular behaviors in the medial part of the lateral plate mesoderm (LPM) at the onset of gonadogenesis. (**a**) In the medial LPM (M-LPM) of E2.0 embryos, ventral (purple) and dorsal (pale pink) cells have different characteristics involved in correct initiation of gonadogenesis. (**b**) Ventral M-LPM cells initiate gonadogenesis by undergoing ingression at E2. They attract primordial germ cells (PGCs), which appear at embryonic bodies from the blood vessel, until E3.0. By contrast, dorsal M-LPM cells remain as epithelium and do not gather PGCs. (**c**) Ventral M-LPM-derived cells express gonadal markers such as GATA4 and form bipotential gonads, whereas dorsal M-LPM-derived cells form the mesonephric capsule (MN), which covers the mesonephros in E4.5 embryos. *DA* dorsal aorta

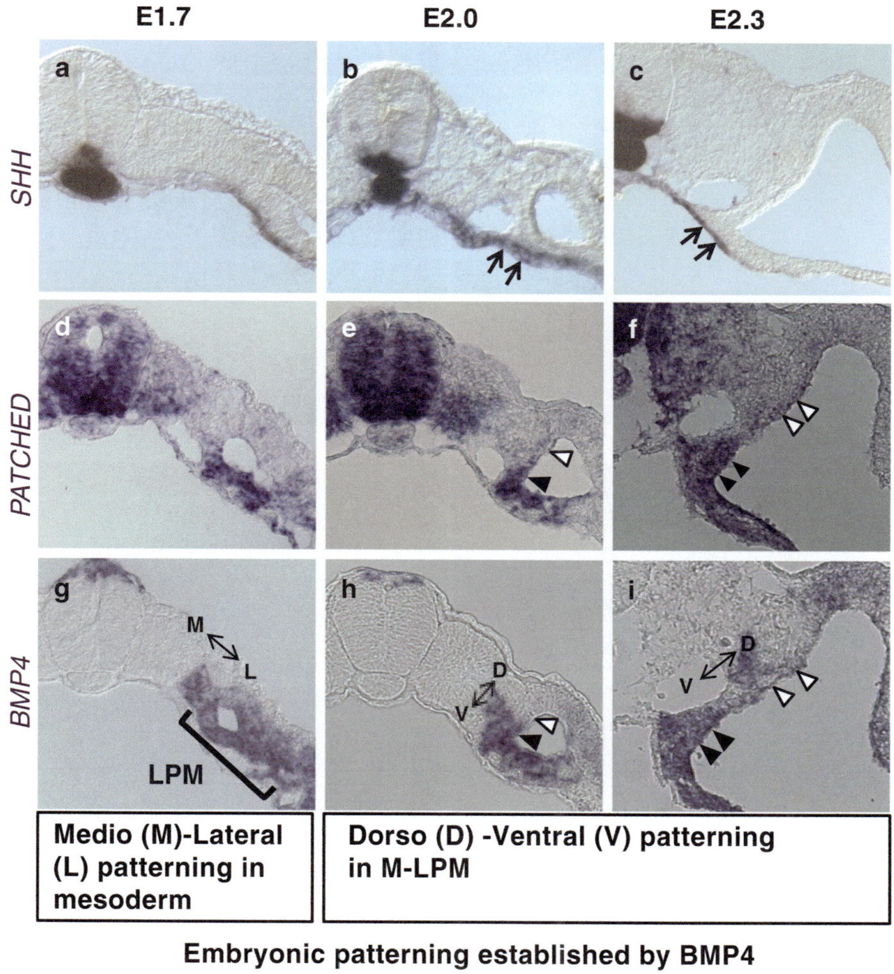

Fig. 2.2 Dorsoventral patterning of M-LPM in E2 embryo is established by Hh and BMP4 signaling. (**a–f**) Expression of *SHH* (**a–c**) and *PATCHED*, representing the activation of Hh signaling (**d–f**), in E2 embryos. *SHH* is expressed in endoderm (black arrows). SHH protein seems to reach, and activate downstream signaling in, ventral (black arrowheads) but not dorsal M-LPM (white arrowheads). (**g–i**) Dynamic transition of BMP4 expression in E2 embryo. Before E2.0, BMP4 is expressed in all LPM cells, and its expression exhibits medio (M)-lateral (L) patterning in mesoderm (**g**). After E2.0, BMP4 expression is repressed in the dorsal M-LPM (white arrowheads) but is maintained in the ventral M-LPM (black arrowheads), thus establishing dorso (D)-ventral (V) patterning in the M-LPM (**h** and **i**)

at this area by overexpressing a secretory Hh inhibitor, HipΔC22 (Fig. 2.3a) [25]. As a result, ventral M-LPM-derived cells could not express the gonadal marker GATA4. This result indicates that ventral M-LPM cells require Hh signaling in order to initiate gonadal differentiation (Fig. 2.3b, d, e, g). Furthermore, they could not form the ridge structure, which seems to be formed by ingression of ventral

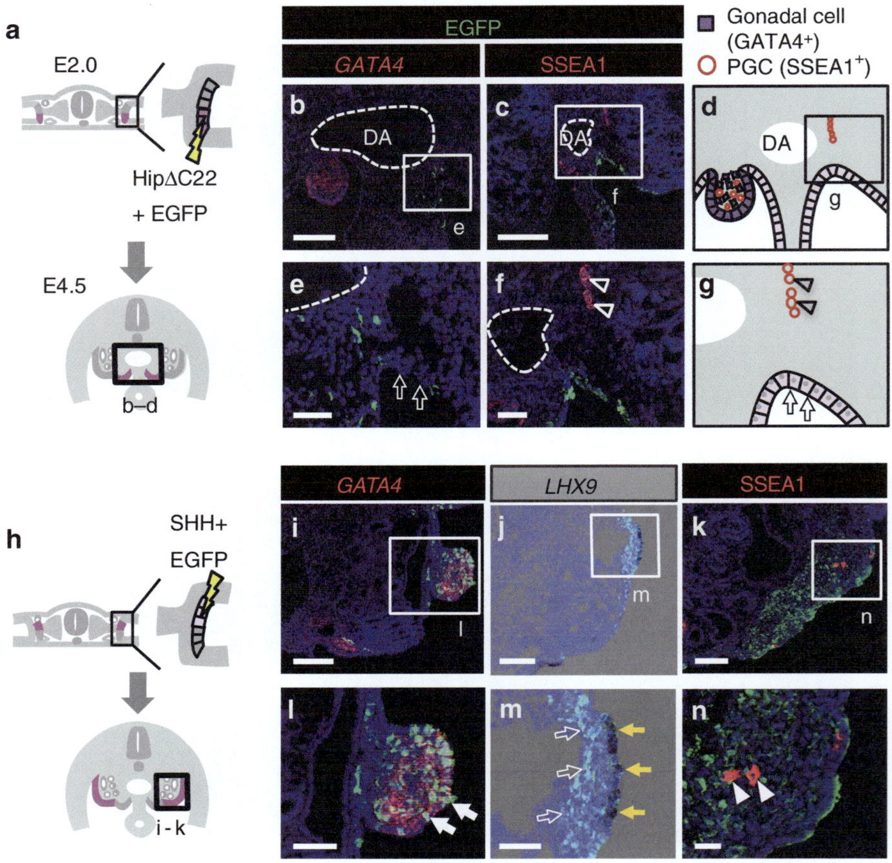

Fig. 2.3 Hh signaling confers ventral character and triggers gonadogenesis in the M-LPM. (**a**) A plasmid-encoding HipΔC22, a secreted Hh signaling inhibitor, was electroporated into the ventral M-LPM at E2.0 along with an EGFP-expressing plasmid. (**b–d**) Boxed region indicated in A in E4.5 embryos, showing *GATA4* mRNA (**b**) and the localization of SSEA1$^+$ PGCs (**c**). A schematic representation of **b** and **c** is shown in (**d**). The bipotential gonad, which is detected as a GATA4$^+$ ridge structure, is formed on the non-electroporated control left side (**b**, **d**). This bipotential gonad retains PGCs (**c**, **d**). (**e–g**) Magnified views of the boxed region in **b–d**. When Hh signaling was inhibited, ventral M-LPM cells could not form bipotential gonad, which appears as a GATA4$^+$ ridge (open arrows in **e** and **g**). In addition, they could not gather PGCs, and PGCs were scattered (open arrowheads in **f** and **g**). (**h**) *SHH* cDNA and *EGFP* cDNA were electroporated into dorsal M-LPM of E2.0 embryos. (**e–k**) Boxed region indicated in **h** in a SHH-electroporated embryo, showing *GATA4* mRNA (**i**), *LHX9* mRNA (**j**), and SSEA1$^+$ PGCs (**k**). Activation of Hh signaling causes dorsal M-LPM cells to ectopically form a GATA4-expressing genital ridge at the surface of the mesonephros (arrows in **l**). This ectopic gonad consists of LHX9$^+$ gonadal coelomic epithelium (yellow arrows in M) and LHX9$^-$ underlying mesenchyme (open arrows in **m**) and retains PGCs (white arrowheads in **n**). Scale bars, 100 μm (**b**, **c**, **i–k**); 50 μm (**e**, **f**, **m**, **n**); 25 μm (**l**)

M-LPM cells. In addition, Hh signaling confers on ventral M-LPM cells the capacity to gather and retain PGCs. When HipΔC22 is overexpressed in ventral M-LPM cells, PGCs were not retained in the bipotential gonad and were instead scattered throughout the embryonic body (Fig. 2.3c, d, f, g). Collectively, Hh signaling activated in the ventral part of the M-LPM is essential for a variety of cellular behaviors involved in initiating gonadogenesis.

By contrast, Hh signaling is sufficient to trigger gonadogenesis in the dorsal M-LPM, which covers the mesonephros. When Hh signaling in dorsal M-LPM cells is ectopically activated by electroporation of SHH ligand, these cells undergo ingression like ventral M-LPM cells. By contrast, control cells remain as epithelium. Moreover, dorsal M-LPM-derived cells form a beautiful genital ridge-like structure and express the gonadal marker GATA4 over the mesonephros upon ectopic activation of Hh signaling (Fig. 2.3h, i, l). This ectopic gonad seems to contain the important structures of the endogenous gonad. The endogenous bipotential gonad consists of stem-like outer coelomic epithelium (G-CE) and inner mesenchyme [12]. At later stages, the G-CE becomes the OSE, which includes ovarian stem cells [3]. Therefore, correct formation of early gonadal structure serves as the basis of the adult ovarian structure. LHX9, which is expressed in the OSE [26], is also specifically expressed in the overlying G-CE in the early embryo [12, 27]. Interestingly, even in ectopic gonad, cells that remain as epithelium express LHX9, whereas cells that undergo ingression and become underlying mesenchyme do not (Fig. 2.3j, m). Therefore, it seems that Hh signaling induces stem-like G-CE at the appropriate location. In addition, SSEA1^{+} PGCs are gathered and retained at this ectopic gonad covering the mesonephros (Fig. 2.3k, n). These results indicate that Hh signaling is sufficient to trigger a variety of cellular behaviors, resulting in formation of a well-organized bipotential gonad in the dorsal M-LPM. Taken together, we found that dorsoventral patterning of the M-LPM, which is established by Hh signaling, plays critical roles in the correct initiation of gonadogenesis (Fig. 2.4a).

2.1.3 Dynamic Transition of BMP4 Expression and Roles at the Downstream of Hh Signaling in LPM

In addition, we revealed an important concept regarding the transition of early embryonic patterning established by BMP4. BMP4, a secretory ligand that causes various cellular behaviors during early embryogenesis and carcinogenesis [28–30], is upregulated by Hh signaling in several tissues (e.g., gut mesenchyme) [31]. Before the initiation of gonadogenesis in the early embryo, such as E1.7, BMP4 is expressed in the whole LPM (Fig. 2.2g). As a result, a BMP4 protein gradient forms from lateral to medial, and this gradient establishes mediolateral patterning in the mesoderm. This mediolateral patterning must determine the position of the M-LPM [29, 32, 33]. Importantly, after the onset of gonadogenesis (E2), BMP4 expression is immediately downregulated in the dorsal part but is maintained in the ventral part of the M-LPM (Fig. 2.2h, i). This dorsoventral BMP4 expression pattern in M-LPM is established by Hh signaling. When Hh signaling is blocked by overexpression of HipΔC22 in the ventral M-LPM of the E2 embryo, BMP4 expression is rapidly downregulated. By contrast, ectopic activation of Hh signaling by electroporation of SHH causes dorsal M-LPM cells to maintain BMP4 expression even after E2. In

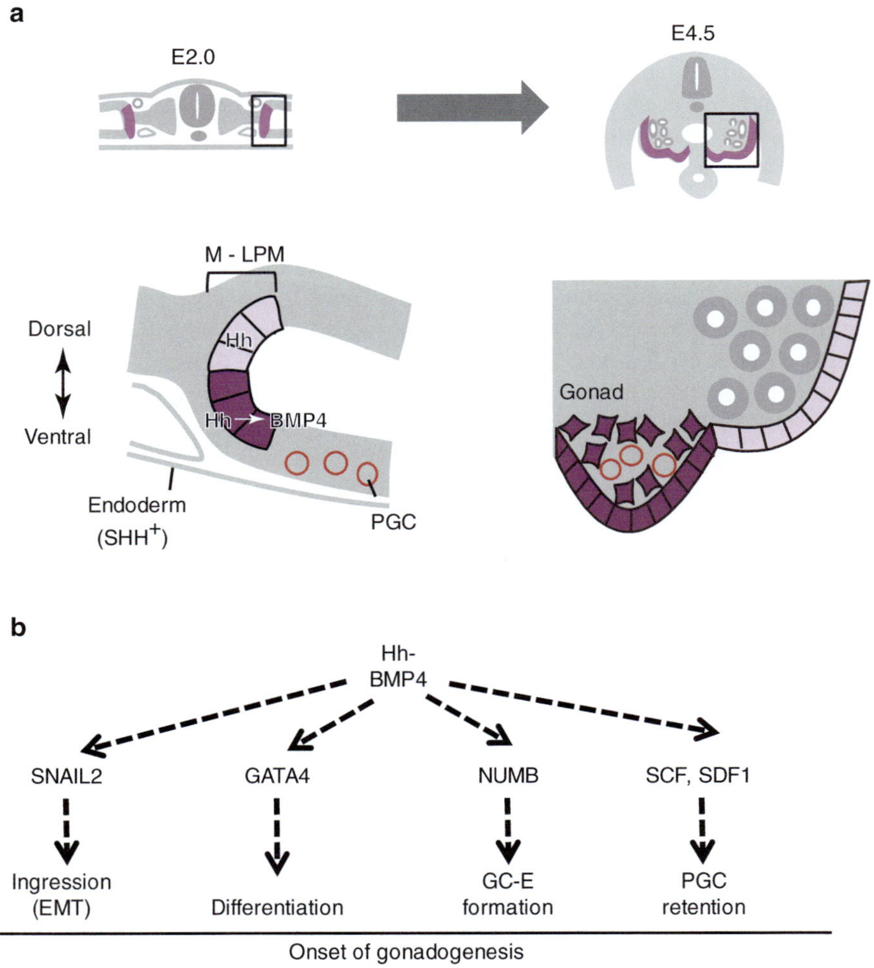

Fig. 2.4 A model of the molecular cascade triggering the gonadogenesis. (**a**) Hh-BMP4 signaling establishes dorsoventral patterning and initiates gonadogenesis in M-LPM of chicken embryo. In M-LPM of E2.0 embryos, Hh signaling is activated in the ventral part (purple) but not in dorsal part (pale pink). This Hh signaling maintains BMP4 expression in ventral M-LPM. As a result, ventral M-LPM cells trigger gonadogenesis by undergoing ingression, differentiating into gonadal cells, and creating PGC niche activity. SHH, a Hedgehog ligand, is expressed in the endoderm, adjacent to the ventral M-LPM. (**b**) Candidate molecules that regulate multiple cellular behaviors involved in the initiation of gonadogenesis downstream of Hh-BMP4 signaling. SNAIL2 is thought to be upregulated in order to induce the epithelial-to-mesenchymal transition (EMT) and ingression, whereas GATA4 expression is induced to promote gonadal differentiation and further development. In mouse embryos, NUMB plays essential roles in the correct formation of stem-like G-CE, whereas expression of SCF and SDF1 is induced by BMP signaling to promote PGC retention. Hh-BMP signaling might orchestrate the expression of these molecules to trigger multiple cellular behaviors in the correct manner

summary, before E2, BMP4 is expressed in the whole LPM, but subsequently a dorsoventral BMP4 expression pattern forms in a Hh signaling-dependent manner.

This novel BMP4 expression pattern after E2 plays pivotal roles in the initiation of gonadogenesis. When BMP signaling is continually activated in dorsal M-LPM even after E2 by electroporation of BMP4, dorsal M-LPM cells exhibit various behaviors related to initiation of gonadogenesis; e.g., they undergo ingression like gonadal progenitor cells and ectopically express the gonadal marker GATA4. Importantly, these cells become able to collect and retain SSEA1^{+} PGCs in their vicinity. Furthermore, BMP4 is sufficient to induce well-orchestrated stem-like G-CE. When BMP4 is overexpressed in dorsal M-LPM, the overlying coelomic epithelial cells ectopically express the G-CE marker LHX9, whereas the underlying mesenchymal cells do not. However, overexpression of BMP4 does not cause dorsal M-LPM cells to form the ridge structure, whereas Hh signaling does. Therefore, it seems that formation of the ridge structure is performed by other mechanisms downstream of Hh signaling.

By contrast, BMP signaling, which is activated in ventral M-LPM cells, is required to trigger gonadogenesis. When BMP signaling is inhibited in ventral M-LPM by electroporation of the secretory BMP inhibitor noggin after E2, these cells do not express the gonadal marker GATA4, do not form the ridge structure, and cannot gather.

PGCs. Collectively, these findings demonstrate that BMP4, whose expression establishes the dorsoventral pattern in the M-LPM (ventral+, dorsal−) after E2, plays central roles in triggering a variety of cellular behaviors involved in onset of gonadogenesis (Fig. 2.4a). It is important to determine how BMP4, which establishes the mediolateral axis in the mesoderm before E2, changes its role to establish the dorsoventral axis in the M-LPM after E2.

2.1.4 Candidate Molecules Regulating Cellular Behaviors Involved in the Onset of Gonadogenesis Downstream of Hh-BMP4 Signaling

Next, how does Hh-BMP4 signaling cause the following cellular behaviors involved in the onset of gonadogenesis: (1) EMT and ingression, (2) gonadal differentiation, (3) stem-like G-CE induction, and (4) acquisition of the capability to retain PGCs? Several recent findings provided hints regarding candidate regulators (Fig. 2.4b).

First, EMT and ingression are expected to be activated by SNAIL2 and the small G-protein RhoA. These molecules play central roles in the EMT in several cell types, including cancer cells [34–36]. They also mediate the EMT in a part of the LPM, which forms mesenterial mesenchyme, upon activation by BMP4 [37]. The same mechanisms might also work in gonadal progenitor cells (ventral M-LPM cells), because progenitor cells of gonad and mesenterial mesenchyme appear in almost the same area and are expected to respond to BMP4 signaling in a similar manner.

Second, SMAD4, an effector of BMP signaling, might play pivotal roles in the gonadal differentiation by regulating GATA4 expression. Gonadal differentiation is represented by the expression of several gonad-specific genes, which promote further development of the gonad [38–40]. In particular, expression of GATA4 is a crucial step in this process because it is the earliest marker of the gonad, and GATA4-deficient mice fail to express other gonadogenesis-related genes [41]. However, the molecular mechanisms that induce GATA4 expression in gonadal primordium remain to be delineated. SMAD4 is required for GATA4 expression in the formation of the heart, which is also derived from the LPM [42, 43]. During heart formation, SMAD4 upregulates GATA4 expression by increasing histone H3 acetylation levels in the *GATA4* promoter region [44]. A similar mechanism might play a central role in gonadal differentiation.

On the other hand, Hh-BMP signaling might regulate a Notch signaling inhibitor, NUMB, to induce stem-like G-CE cells. These cells, which express LHX9 [12, 27], maintain a stem-like state at the outer layer of the ovary and give rise to differentiated cells, such as granulosa cells, after ingression. Recent work showed that NUMB plays central roles in these behaviors by regulating asymmetric cell division in mouse embryos. After cell division in the G-CE, one daughter cell remains in the G-CE and maintains a stem-like state, whereas the other daughter cell undergoes ingression and loses this property. NUMB inhibits Notch signaling, which is required for maintenance of the stem-like state, in ingressed cells. By contrast, in the other daughter cell, which remains in the G-CE, NUMB is distributed on the basal side and does not inhibit Notch signaling. Hh-BMP signaling might regulate this mechanism to induce the G-CE.

Furthermore, Hh-BMP4 signaling might upregulate SCF and SDF1 to provide the ability to retain PGCs in gonadal progenitor cells. Many studies have demonstrated that SCF and SDF1, which are expressed in gonadal cells, play pivotal roles in attraction and maintenance of PGCs [13, 45, 46]. However, the mechanisms responsible for upregulating these molecules are poorly understood. One of the BMP receptors, BMPR1A, seems to be required for expression of these molecules in mouse embryos [47]. Hh-BMP4 signaling might activate BMPR1A to upregulate SCF and SDF1 and enable gonadal cells to retain PGCs.

Collectively, these observations suggest that Hh-BMP4 signaling elicits the expression of several molecules, such as SNAIL2, GATA4, NUMB, and SDF1, in parallel to trigger gonadogenesis in ventral M-LPM cells (Fig. 2.4b). Future work should seek to uncover how expression of these molecules is orchestrated by Hh-BMP4 signaling.

Conclusion

The fundamental characteristics of the ovary are established in the embryonic bipotential gonad. Formation of this embryonic tissue is triggered by ingression of a part of coelomic lining cells juxtaposed to the mesentery. Subsequently, these cells exhibit a variety of behaviors, ultimately leading to formation of a functional gonad. Although all organogenesis is initiated based on early embryonic patterning, it remains poorly understood what kind of embryonic patterning regulates these behaviors.

We recently addressed this long-standing question by using chicken embryos. Gonadogenesis seems to be triggered in early embryos (chicken E2 and mouse E8.5), in which multiple types of pattern formation occur, and the tissue of the coelomic lining is called the LPM. At this stage, cells in the ventral M-LPM start ingression and initiate other behaviors involved in correct formation of the gonad, whereas cells in the dorsal M-LPM do not.

These contrasting behaviors are caused by dorsoventral patterning, which is established by Hh-BMP4 signaling in the M-LPM. First, SHH, which is secreted from endoderm, reaches and activates its downstream Hh signaling in ventral but not dorsal M-LPM. Consequently, BMP4 seems to change its expression and role in LPM. Before the initiation of gonadogenesis, BMP4 is secreted from all LPM cells and serves to establish the mediolateral axis in the mesoderm. However, at the onset of gonadogenesis, BMP4 expression is restricted to the ventral part in M-LPM in a Hh signaling-dependent manner. Furthermore, BMP4 plays a distinct role in orchestrating various behaviors involved in the initiation of gonadogenesis.

Hh-BMP signaling might upregulate several molecules, such as SNAIL2, NUMB, GATA4, SCF, and SDF1, in ventral M-LPM in order to control the complex cellular behaviors at the onset of gonadogenesis. In future studies, it will be necessary to comprehensively study the complicated molecular cascades downstream of Hh-BMP4 signaling that organize the initiation of gonadogenesis.

References

1. Li R, Albertini DF. The road to maturation: somatic cell interaction and self-organization of the mammalian oocyte. Nat Rev Mol Cell Biol. 2013;14:141–52.
2. Young JM, McNeilly AS. Theca: the forgotten cell of the ovarian follicle. Reproduction (Cambridge, England). 2010;140:489–504.
3. Auersperg N, Wong AS, Choi KC, Kang SK, Leung PC. Ovarian surface epithelium: biology, endocrinology, and pathology. Endocr Rev. 2001;22:255–88.
4. Katabuchi H, Okamura H. Cell biology of human ovarian surface epithelial cells and ovarian carcinogenesis. Med Electr Microsc. 2003;36:74–86.
5. Koopman P. The curious world of gonadal development in mammals. Curr Top Dev Biol. 2016;116:537–45.
6. Suzuki H, Kanai-Azuma M, Kanai Y. From sex determination to initial folliculogenesis in mammalian ovaries: morphogenetic waves along the anteroposterior and dorsoventral axes. Sex Dev. 2015;9:190–204.
7. Biason-Lauber A, Chaboissier MC. Ovarian development and disease: the known and the unexpected. Semin Cell Dev Biol. 2015;45:59–67.
8. Harikae K, Miura K, Kanai Y. Early gonadogenesis in mammals: significance of long and narrow gonadal structure. Dev Dyn. 2013;242:330–8.
9. DeFalco T, Capel B. Gonad morphogenesis in vertebrates: divergent means to a convergent end. Annu Rev Cell Dev Biol. 2009;25:457–82.
10. Piprek RP, Kloc M, Kubiak JZ. Early development of the gonads: origin and differentiation of the somatic cells of the genital ridges. Results Probl Cell Differ. 2016;58:1–22.
11. Tanaka SS, Nishinakamura R. Regulation of male sex determination: genital ridge formation and Sry activation in mice. Cell Mol Life Sci. 2014;71:4781–802.

12. Lin YT, Barske L, DeFalco T, Capel B. Numb regulates somatic cell lineage commitment during early gonadogenesis in mice. Development (Cambridge, England). 2017;144:1607–18.
13. Richardson BE, Lehmann R. Mechanisms guiding primordial germ cell migration: strategies from different organisms. Nat Rev Mol Cell Biol. 2010;11:37–49.
14. Yoshino T, Murai H, Saito D. Hedgehog-BMP signalling establishes dorsoventral patterning in lateral plate mesoderm to trigger gonadogenesis in chicken embryos. Nat Commun. 2016;7:12561.
15. Atsuta Y, Takahashi Y. FGF8 coordinates tissue elongation and cell epithelialization during early kidney tubulogenesis. Development (Cambridge, England). 2015;142:2329–37.
16. Saito D, Takase Y, Murai H, Takahashi Y. The dorsal aorta initiates a molecular cascade that instructs sympatho-adrenal specification. Science (New York, NY). 2012;336:1578–81.
17. Pietila I, Vainio S. The embryonic aorta-gonad-mesonephros region as a generator of haematopoietic stem cells. APMIS. 2005;113:804–12.
18. de Bruijn MF, Speck NA, Peeters MC, Dzierzak E. Definitive hematopoietic stem cells first develop within the major arterial regions of the mouse embryo. EMBO J. 2000;19:2465–74.
19. Yoshino T, Saito D, Tadokoro R, Takahashi Y. In vivo gene manipulations of epithelial cell sheets: a novel model to study epithelial-to-mesenchymal transition. Develop Growth Differ. 2011;53:378–88.
20. Yoshino T, et al. Interepithelial signaling with nephric duct is required for the formation of overlying coelomic epithelial cell sheet. Proc Natl Acad Sci U S Am. 2014;111:6660–5.
21. Hen G, Friedman-Einat M, Sela-Donenfeld D. Primordial germ cells in the dorsal mesentery of the chicken embryo demonstrate left-right asymmetry and polarized distribution of the EMA1 epitope. J Anat. 2014;224:556–63.
22. Briscoe J, Small S. Morphogen rules: design principles of gradient-mediated embryo patterning. Development (Cambridge, England). 2015;142:3996–4009.
23. Ramsbottom SA, Pownall ME. Regulation of Hedgehog signalling inside and outside the cell. J Dev Biol. 2016;4:23.
24. Zhang XM, Ramalho-Santos M, McMahon AP. Smoothened mutants reveal redundant roles for Shh and Ihh signaling including regulation of L/R asymmetry by the mouse node. Cell. 2001;105:781–92.
25. Madison BB, et al. Epithelial hedgehog signals pattern the intestinal crypt-villus axis. Development (Cambridge, England). 2005;132:279–89.
26. Auersperg N. The stem-cell profile of ovarian surface epithelium is reproduced in the oviductal fimbriae, with increased stem-cell marker density in distal parts of the fimbriae. Int J Gynecol Pathol. 2013;32:444–53.
27. Oreal E, Mazaud S, Picard JY, Magre S, Carre-Eusebe D. Different patterns of anti-Mullerian hormone expression, as related to DMRT1, SF-1, WT1, GATA-4, Wnt-4, and Lhx9 expression, in the chick differentiating gonads. Dev Dyn. 2002;225:221–32.
28. Kallioniemi A. Bone morphogenetic protein 4-a fascinating regulator of cancer cell behavior. Cancer Genet. 2012;205:267–77.
29. De Robertis EM, Kuroda H. Dorsal-ventral patterning and neural induction in Xenopus embryos. Annu Rev Cell Dev Biol. 2004;20:285–308.
30. Stern CD. Neural induction: old problem, new findings, yet more questions. Development (Cambridge, England). 2005;132:2007–21.
31. Sukegawa A, et al. The concentric structure of the developing gut is regulated by Sonic hedgehog derived from endodermal epithelium. Development (Cambridge, England). 2000;127:1971–80.
32. Tonegawa A, Takahashi Y. Somitogenesis controlled by Noggin. Dev Biol. 1998;202:172–82.
33. Tonegawa A, Funayama N, Ueno N, Takahashi Y. Mesodermal subdivision along the mediolateral axis in chicken controlled by different concentrations of BMP-4. Development (Cambridge, England). 1997;124:1975–84.
34. Nieto MA, Huang RY, Jackson RA, Thiery JP. EMT: 2016. Cell. 2016;166:21–45.
35. Nakaya Y, Sheng G. An amicable separation: Chick's way of doing EMT. Cell Adhes Migr. 2009;3:160–3.

36. Nieto MA. Epithelial plasticity: a common theme in embryonic and cancer cells. Science (New York, NY). 2013;342:1234850.
37. Arraf AA, Yelin R, Reshef I, Kispert A, Schultheiss TM. Establishment of the visceral embryonic midline is a dynamic process that requires bilaterally symmetric BMP signaling. Dev Cell. 2016;37:571–80.
38. Morohashi K. The ontogenesis of the steroidogenic tissues. Genes Cells. 1997;2:95–106.
39. Ozisik G, Achermann JC, Meeks JJ, Jameson JL. SF1 in the development of the adrenal gland and gonads. Horm Res. 2003;59:94–8.
40. Birk OS, et al. The LIM homeobox gene Lhx9 is essential for mouse gonad formation. Nature. 2000;403:909–13.
41. Hu YC, Okumura LM, Page DC. Gata4 is required for formation of the genital ridge in mice. PLoS Genet. 2013;9:e1003629.
42. Qi X, et al. Essential role of Smad4 in maintaining cardiomyocyte proliferation during murine embryonic heart development. Dev Biol. 2007;311:136–46.
43. Rojas A, et al. Gata4 expression in lateral mesoderm is downstream of BMP4 and is activated directly by Forkhead and GATA transcription factors through a distal enhancer element. Development (Cambridge, England). 2005;132:3405–17.
44. Si L, et al. Smad4 mediated BMP2 signal is essential for the regulation of GATA4 and Nkx2.5 by affecting the histone H3 acetylation in H9c2 cells. Biochem Biophys Res Commun. 2014;450:81–6.
45. Mamsen LS, Brochner CB, Byskov AG, Mollgard K. The migration and loss of human primordial germ stem cells from the hind gut epithelium towards the gonadal ridge. Int J Dev Biol. 2012;56:771–8.
46. Stebler J, et al. Primordial germ cell migration in the chick and mouse embryo: the role of the chemokine SDF-1/CXCL12. Dev Biol. 2004;272:351–61.
47. Dudley B, Palumbo C, Nalepka J, Molyneaux K. BMP signaling controls formation of a primordial germ cell niche within the early genital ridges. Dev Biol. 2010;343:84–93.

Regulation of Metabolic Pathways in Steroidogenic Cells by Ad4BP/SF-1

3

Ken-ichirou Morohashi, Miki Inoue, Bing Li, and Takashi Baba

Abstract

Ad4BP/SF-1 (NR5A1), a member of the nuclear receptor superfamily, is known to play crucial roles in the regulation of steroidogenesis in the gonads and adrenal cortex, and many studies have demonstrated that all steroidogenic genes are direct targets of Ad4BP/SF-1. In addition, in vivo KO studies demonstrated that no steroidogenic organs formed in gene-disrupted mice, strongly suggesting that Ad4BP/SF-1 is essential for the development of the gonads and adrenal gland. However, it remains unclear how Ad4BP/SF-1 regulates the development of the steroidogenic organs and which additional non-steroidogenic genes are targeted.

We surveyed the target genes of Ad4BP/SF-1 in steroidogenic cells by mRNA deep sequencing and ChIP-sequence analyses using an Ad4BP/SF-1 antibody. Nearly all genes in energy-producing glycolytic pathways were found to be regulated by Ad4BP/SF-1. We also showed that the key genes implicated in NADPH production are the direct targets. Since sufficient supplies of ATP and NADPH are necessary for cellular survival and production of steroids, Ad4BP/SF-1 possibly orchestrates tissue-specific steroidogenic pathway and these housekeeping metabolic pathways.

Keywords

Steroidogenic cells · Glycolysis · NADPH synthesis · Ad4BP/SF-1

K.-i. Morohashi (✉) · T. Baba
Department of Molecular Biology, Graduate School of Medical Sciences, Kyushu University, Fukuoka, Japan

Division of Molecular Life Science, Graduate School of Systems Life Science, Kyushu University, Fukuoka, Japan
e-mail: moro@cell.med.kyushu-u.ac.jp

M. Inoue · B. Li
Division of Molecular Life Science, Graduate School of Systems Life Science, Kyushu University, Fukuoka, Japan

H. Katabuchi et al. (eds.), *Cell Biology of the Ovary*,
https://doi.org/10.1007/978-981-10-7941-2_3

3.1 Introduction

A member of the nuclear receptor superfamily, Ad4BP/SF-1 (NR5A1), is expressed in the steroidogenic adrenal cortex and gonads [1–5]. This protein was originally identified as a transcription factor regulating the steroidogenic *CYP11A1* and *CYP11B1* genes [6, 7]. To date, many studies have established that Ad4BP/SF-1 controls steroid hormone synthesis through the transcriptional regulation of all steroidogenic genes. In addition to steroidogenic regulation, Ad4BP/SF-1 plays a crucial role in steroidogenic tissue development as revealed by the gene knockout mice. Namely, those animals lacked steroidogenic adrenal glands and gonads [8–11]. By contrast, the overexpression of *Ad4BP/SF-1* induced enlargement/ectopic formation and tumorigenesis of the adrenal cortex [12–15]. Taken together, these studies strongly suggested that Ad4BP/SF-1 is involved in the regulation of cellular proliferation and maintenance. However, it is unlikely that Ad4BP/SF-1 regulates these biological events through steroidogenic gene regulation.

The production of important coenzymes for energy potential and reducing power, e.g., ATP and NADPH, is mediated by nutrient catabolism. Glucose, the representative nutrient, is catabolized through the glycolytic pathway and subsequently by the TCA cycle and oxidative phosphorylation. This sequential process ultimately produces ATP, which thereafter serves as the energy source for many metabolic processes. NADPH is a crucial source of power in reductive reactions in a variety of cellular biosynthetic reactions, such as the synthesis of fatty acids, cholesterol, and steroid hormones. In addition to these reactions, NADPH is consumed during the metabolization of xenobiotic (toxic) compounds (including drugs), maintenance of reduced glutathione (GSH) pools, and production of superoxide and reactive oxygen species [16, 17]. Considering its involvement in these pivotal functions, NADPH is regarded as a critical compound for cellular activity, proliferation, and survival [18].

3.1.1 Metabolic Pathways Activated in Steroidogenic Cells

Steroidogenesis is activated in steroidogenic cells, such as the adrenal cortex, testicular Leydig cells, and ovarian theca and granulosa cells. Glucocorticoid and mineralocorticoid are synthesized in the adrenal cortex, while the sex steroids, testosterone and estradiol, are synthesized in the testes and ovary, respectively. Steroid hormones are synthesized from cholesterol via multiple enzymatic reactions [19]. The first step in the pathway is the conversion of cholesterol to pregnenolone, which is mediated by CYP11A1 (encoding cholesterol side-chain cleavage P450) localized in mitochondria. In addition to CYP11A1, five forms of CYPs and two forms of hydroxysteroid dehydrogenase are involved in steroidogenesis.

Nicotinamide adenine dinucleotide phosphate (NADPH) is a crucial cofactor that supplies reductive power to a variety of cellular biosynthetic reactions, such as the synthesis of fatty acids and cholesterol. NADPH is also consumed during steroidogenesis in reactions mediated by steroidogenic CYPs. Due to its involvement in these pivotal functions, NADPH is regarded as a critical compound for cellular activity, proliferation, and survival [18]. The cellular NADPH pool is maintained through multiple reactions. Of the enzymes involved in the pentose phosphate pathway, glucose-6-phosphate dehydrogenase and 6-phosphogluconate dehydrogenase produce NADPH. Isocitrate dehydrogenase (IDH), malic enzyme (ME), and methylenetetrahydrofolate dehydrogenase (MTHFD) also mediate NADPH production [20–22].

Ad4BP/SF-1 (NR5A1) is a member of the nuclear receptor superfamily and is expressed specifically in the steroidogenic adrenal cortex and gonads [2, 23, 24]. This protein was originally identified as a transcription factor regulating the steroidogenic *CYP11A1* and *CYP11B1* encoding steroid 11β-hydroxylase P450 [25]. Subsequent studies eventually demonstrated that all steroidogenic genes are targets of Ad4BP/SF-1. To date, a collective body of work has established that Ad4BP/SF-1 controls steroid hormone synthesis through the transcriptional regulation of all steroidogenic genes [19]. In addition to steroidogenic regulation, Ad4BP/SF-1 plays a crucial role in steroidogenic tissue development as shown in mice bearing targeted deletions in the gene; those animals lacked steroidogenic organs such as adrenal glands and gonads [8–10]. In contrast, the overexpression of *Ad4BP/SF-1* induced enlargement/ectopic formation and tumorigenesis of the adrenal cortex [12–15]. Taken together, these studies strongly suggested that Ad4BP/SF-1 is involved in the regulation of cellular proliferation and maintenance. However, it is unlikely that Ad4BP/SF-1 controls cellular proliferation and maintenance through steroidogenic gene regulation. The aim of our study was therefore to investigate the mechanisms involved in the regulation of such a variety of cellular events by Ad4BP/SF-1.

3.1.2 Implication of Ad4BP/SF-1 into Glycolytic Gene Regulation

To determine all the target genes of Ad4BP/SF-1, transcriptomes were obtained from Y-1 cells derived from mouse adrenocortical tumor and MA-10 mouse Leydig cells. Adult and fetal types of testicular Leydig cells prepared from adult and fetal mouse testes, respectively, were also analyzed, as these cells are positive for Ad4BP/SF-1 and thus exhibit steroidogenic activities. By comparing the transcriptomes obtained from the control and Ad4BP/SF-1 knockdown (KD) cells, we found that the expression of many genes was up- or downregulated in Ad4BP/SF-1 KD cells. This provided candidate genes that were positively regulated by Ad4BP/SF-1 [26].

Gene ontology analysis of the downregulated genes suggested that Ad4BP/SF-1 may be involved in a variety of metabolic processes (Table 3.1). Interestingly, many of the processes included the glycolytic pathway. In fact, in addition to the steroidogenic genes previously established as targets of Ad4BP/SF-1, the expression of all glycolytic genes (*Hk1*, *Gpi1*, *Pfkl*, *Pfkp*, *Aldoa*, *Tpi1 Gapdh*, *Pgk1*, *Pgam1*, *Eno1*, and *Pkm2*) was decreased in the KD cells (Fig. 3.1). The decreased expression of these genes was validated by quantitative RT-PCR (qRT-PCR). Downregulation of the glycolytic genes was observed in the KD studies with MA-10 and adult Leydig cells, suggesting that Ad4BP/SF-1 is involved in glycolytic gene regulation of all

Table 3.1 Glycolytic genes involved in a variety of pathways affected by siAd4BP/SF-1 treatment

P-Value	Terms
4.23 *E*–07	Organic acid metabolic process
4.23 *E*–07	Carboxylic acid metabolic process
4.23 *E*–07	Oxoacid metabolic process
5.98 *E*–07	Cellular ketone metabolic process
1.38 *E*–06	Alcohole catabolic process
4.21 *E*–06	Regulation of biological quality
5.73 *E*–06	Glucose catabolic process
5.73 *E*–06	Hexose catabolic process
6.37 *E*–06	Monosaccharide catabolic process
1.10 *E*–05	Small molecule metabolic process
1.34 *E*–05	Monocarboxylic acid metabolic process
2.60 *E*–05	Glycolysis
2.75 *E*–05	Cellular carbohydrate catabolic process
4.63 *E*–05	Glucose metabolic process
5.95 *E*–05	Response to organic substance
6.18 *E*–05	Cellular response to chemical substance
6.61 *E*–05	Fat cell differentiation

The 243 genes down-regulated by siAd4BP/SF-1 treatment in Y-1 cells were analyzed using GOstats software

steroidogenic cells. The expression of genes involved in other metabolic pathways, specifically the TCA cycle and fatty acid oxidation, was not coordinately affected by the KD.

To determine whether Ad4BP/SF-1 binds directly to glycolytic genes, ChIP-seq (chromatin immunoprecipitation sequencing) was performed using a specific Ad4BP/SF-1 antibody in Y-1 cells and Leydig cells from adult mouse testes. As expected, Ad4BP/SF-1 was found to accumulate in many glycolytic genes in Y-1 cells. Likewise, such accumulation was observed in adult Leydig cells, supporting the assumption that glycolytic genes are directly regulated by Ad4BP/SF-1 (Fig. 3.1).

Among the glycolytic genes expressed in Y-1 and Leydig cells, Ad4BP/SF-1 binding sites were localized in the upstream and/or intronic regions of nine genes (including the paralogues *Pfkl* and *Pfkp*) that mediate 8 of the 11 glycolytic reactions. The ChIP peaks were observed within 10 kb upstream of the *Eno1*, *Pfkl*, and *Pkm2* genes and within the intronic regions of the *Hk1*, *Pfkp*, *Aldoa*, *Gapdh*, *Pgam1*, *Eno1*, *Pkm2*, and *Ldha* genes (Fig. 3.2). As expected, the Ad4BP/SF-1 binding sequence was localized in those regions. The peak regions were then subjected to luciferase reporter gene assays to determine transcriptional activities and were found to exhibit enhancer activities through the binding of Ad4BP/SF-1. No ChIP peaks were observed in the *Gpi1*, *Tpi1*, and *Pgk1* gene loci, even though their gene expression was downregulated by *Ad4BP/SF-1* KD. These observations strongly suggest that the glycolytic pathway is regulated by Ad4BP/SF-1 at the level of gene expression. Given the roles of Ad4BP/SF-1, it was expected that Ad4BP/SF-1 KD would decrease glycolysis and thus decrease the cellular concentration of ATP. As expected, ATP concentration in *Ad4BP/SF-1* KD Y-1 cells was found to be lowered.

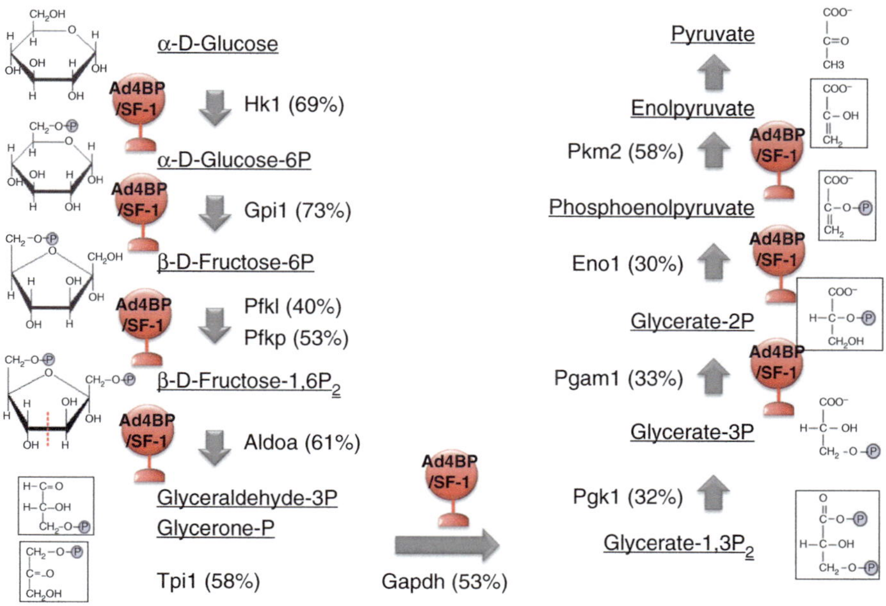

Fig. 3.1 Regulation of the glycolytic pathway by Ad4BP/SF-1 revealed by transcriptome and ChIP-Seq studies. Transcriptomes obtained from control and *Ad4BP/SF-1* KD cells demonstrated that the expression of all glycolytic genes decreased to the level indicated in parentheses compared with control cells. ChIP-seq studies with a specific Ad4BP/SF-1 antibody demonstrated that Ad4BP/SF-1 accumulated in the genes highlighted with a picture representing Ad4BP/SF-1. Together, these results strongly suggest that most, if not all, glycolytic genes are the direct targets of Ad4BP/SF-1

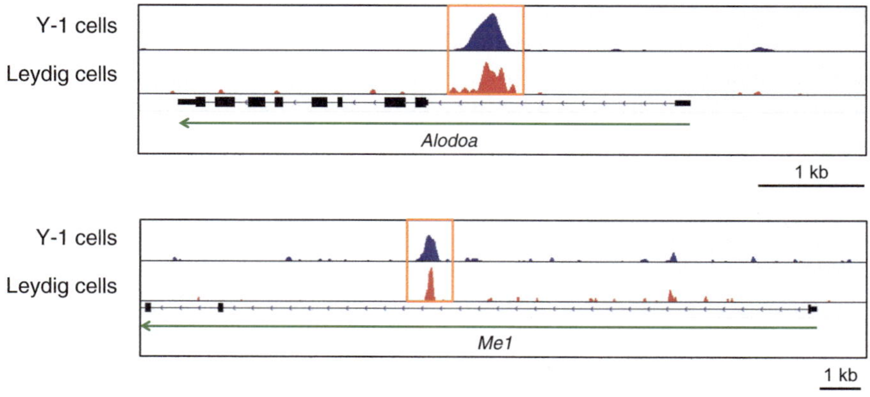

Fig. 3.2 Accumulation of Ad4BP/SF-1 to glycolytic and NADPH synthetic genes. ChIP-seq for Ad4BP/SF-1 was performed using Y-1 and adult testicular Leydig cells. Accumulation of Ad4BP/SF-1 at *Aldos* in the glycolytic pathway (upper) and NADPH synthetic *Me1* gene (lower) is shown. Exons of the genes are shown by black boxes, and Ad4BP/SF-1 peaks are highlighted with orange boxes. Arrows indicate the direction of transcription

This finding is supported by the alteration of gene expression during fetal Leydig differentiation from progenitor cells [27]. To characterize both cell types, we established a mouse line in which fetal Leydig cells are strongly labeled with EGFP while their progenitor cells are labeled very weakly [28, 29]. During the differentiation from progenitor to mature Leydig cells, the expression of *Ad4BP/SF-1* gene was strongly enhanced. The transcriptomes obtained from these cells demonstrated that glycolytic gene expression was strongly activated in mature Leydig cells. Considering the simultaneous increase in Ad4BP/SF-1 and glycolytic gene expression, it can be concluded that glycolytic gene expression is regulated by Ad4BP/SF-1 during the process of fetal Leydig cell differentiation from the progenitors.

3.1.3 Implication of Ad4BP/SF-1 into NADPH Synthetic Gene Regulation

In addition to glycolytic genes, we noticed that the expression of genes involved in NADPH synthesis was decreased. The cellular NADPH pool is maintained through multiple reactions. Of the enzymes involved in the pentose phosphate pathway (PPP), glucose-6-phosphate dehydrogenase (G6PD) and 6-phosphogluconate dehydrogenase (6PGD) produce NADPH. Isocitrate dehydrogenase (IDH), malic enzyme (ME), and methylene tetrahydrofolate dehydrogenase (MTHFD) also mediate NADPH production [20–22].

Ad4BP/SF-1 KD led to a decrease in the expression of *glucose-6-phosphate dehydrogenase* and *6-phosphogluconate dehydrogenase* genes in the pentose phosphate pathway in Y-1 cells [26]. Of the three genes encoding malic enzyme (*Me1*, *Me2*, and *Me3*), the expression of cytosolic *Me1*, which has a higher expression than mitochondrial *Me2* and *Me3*, was decreased by *Ad4BP/SF-1* KD. Of the five genes encoding isocitrate dehydrogenase, cytosolic IDH1 and mitochondrial IDH2 require $NADP^+$ for their enzymatic reaction, while IDH3 (composed of IDH3α, IDH3β, and IDH3γ subunits) requires NAD^+. *Ad4BP/SF-1* KD decreased the expression of *Idh1* but not that of *Idh2*. The expression of cytosolic *Mthfd1* and mitochondrial *Mthfd2* was likewise decreased [30].

As noted above, the genome-wide ChIP-seq study revealed that Ad4BP/SF-1 accumulates in glycolytic genes in adrenocortical Y-1 cells. Since the results discussed above suggest that multiple genes responsible for NADPH synthesis are the possible target genes of Ad4BP/SF-1, we examined the ChIP-seq data sets for relevant peaks. As expected, Ad4BP/SF-1 binding peaks were observed in the *Me1* and *Mthfd2* gene *loci* (Fig. 3.2). Accumulation of Ad4BP/SF-1 in these regions was validated by ChIP-qPCR. Unexpectedly, no ChIP peaks were observed in other genes related to NADPH production, even though their gene expression was decreased by *Ad4BP/SF-1* KD.

NADPH concentration in the total cell lysate was determined in *Ad4BP/SF-1*, *Me1*, and *Mthfd2* KD Y-1 cells. *Mthfd2* KD decreased the cellular NADPH level to approximately 70% of control cells. Likewise, *Me1* KD decreased the level to 80%.

Interestingly, *Ad4BP/SF-1* KD led to a 50% decrease in NADPH, which was larger than that from *Mthfd2* KD or si*Me1* KD. Considering that both *Me1* and *Mthfd2* are possible downstream genes of Ad4BP/SF-1, it can be concluded that NADPH in the *Ad4BP/SF-1* KD cells was lower than in the si*Me1* or si*Mthfd2* KD cells [30].

3.1.4 Orchestration of Metabolic Pathways by a Single Transcription Factor

It is notable that Ad4BP/SF-1 is involved in the regulation of genes necessary for housekeeping cellular activity such as glycolysis and NADPH synthesis, although the factor has previously been established as a tissue-specific transcription factor regulating all steroidogenic gene expression in the gonads and adrenal cortex [9, 24]. Cholesterol is the raw material used in steroid hormone synthesis, supplied to steroidogenic cells through two main routes: incorporation from the blood and de novo synthesis from acetyl-CoA by multiple enzymatic steps. In fact, testicular Leydig cells synthesize a certain percentage of their intracellular cholesterol by themselves [31]. Notably, both ATP and NADPH are consumed in the cholesterol synthetic pathway. In addition, NADPH is required for P450-mediated reactions in the steroid hormone synthetic pathway [32]. Considering the requirement for ATP and NADPH in de novo cholesterogenesis and steroidogenesis from acetyl-CoA, the genes involved in glucose metabolism could be activated cooperatively with the genes involved in steroidogenesis in steroidogenic cells. Moreover, this cooperative regulation of biomass supply and demand for steroidogenesis would be possible if a single key transcription factor were implicated in both pathways. Our results show that Ad4BP/SF-1 is a plausible candidate to orchestrate regulation of tissue-specific steroidogenic genes and housekeeping glucose metabolic genes, thereby enabling tissue-specific functions.

Our study of Ad4BP/SF-1, together with the ERRα study, raises the possibility that a subset of closely related nuclear receptors governs glucose metabolism. In addition, considering that the nucleotide sequence recognized by Ad4BP/SF-1 can be shared by NR5A2 (LRH1) [33], and at least in part by NR4 nuclear receptor family members [34], these nuclear receptors may also share target genes. While Charest-Marcotte et al. referred to the set of genes regulated by ERRα as an "ERRα regulon" [35, 36], our study indicates that a subset of nuclear receptors may regulate metabolic processes by constructing regulons. In addition, our present study is the first to suggest that a single nuclear receptor may govern cellular growth/maintenance and tissue-specific functions by balancing biomass supply and demand.

Conclusion

Although Ad4BP/SF-1 has been established as a tissue-specific transcription factor regulating all steroidogenic gene expression in the gonads and adrenal cortex, our current studies demonstrated that the transcription factor also regulates the genes involved in housekeeping metabolic processes, glycolysis and NADPH production. Considering that the energized molecules, ATP and NADPH, are necessary for cholesterogenesis and steroidogenesis, orchestration of the multiple metabolic pathways by Ad4BP/SF-1 enables steroidogenic cells to efficiently produce steroids.

References

1. Hammer GD, Parker KL, Schimmer BP. Minireview: transcriptional regulation of adrenocortical development. Endocrinology. 2005;146:1018–24.
2. Hatano O, Takayama K, Imai T, Waterman MR, Takakusu A, Omura T, Morohashi K. Sex-dependent expression of a transcription factor, Ad4BP, regulating steroidogenic P-450 genes in the gonads during prenatal and postnatal rat development. Development. 1994;120:2787–97.
3. Hoivik EA, Lewis AE, Aumo L, Bakke M. Molecular aspects of steroidogenic factor 1 (SF-1). Mol Cell Endocrinol. 2010;315:27–39.
4. Morohashi K, Honda S, Inomata Y, Handa H, Omura T. A common trans-acting factor, Ad4-binding protein, to the promoters of steroidogenic P-450s. J Biol Chem. 1992;267:17913–9.
5. Val P, Lefrancois-Martinez AM, Veyssiere G, Martinez A. SF-1 a key player in the development and differentiation of steroidogenic tissues. Nucl Recept. 2003;1:8.
6. Morohashi K, Zanger UM, Honda S, Hara M, Waterman MR, Omura T. Activation of CYP11A and CYP11B gene promoters by the steroidogenic cell-specific transcription factor, Ad4BP. Mol Endocrinol. 1993;7:1196–204.
7. Clemens JW, Lala DS, Parker KL, Richards JS. Steroidogenic factor-1 binding and transcriptional activity of the cholesterol side-chain cleavage promoter in rat granulosa cells. Endocrinology. 1994;134:1499–508.
8. Luo X, Ikeda Y, Parker KL. A cell-specific nuclear receptor is essential for adrenal and gonadal development and sexual differentiation. Cell. 1994;77:481–90.
9. Morohashi KI, Omura T. Ad4BP/SF-1, a transcription factor essential for the transcription of steroidogenic cytochrome P450 genes and for the establishment of the reproductive function. FASEB J. 1996;10:1569–77.
10. Sadovsky Y, Crawford PA, Woodson KG, Polish JA, Clements MA, Tourtellotte LM, Simburger K, Milbrandt J. Mice deficient in the orphan receptor steroidogenic factor 1 lack adrenal glands and gonads but express P450 side-chain-cleavage enzyme in the placenta and have normal embryonic serum levels of corticosteroids. Proc Natl Acad Sci U S A. 1995;92:10939–43.
11. Shinoda K, Lei H, Yoshii H, Nomura M, Nagano M, Shiba H, Sasaki H, Osawa Y, Ninomiya Y, Niwa O, et al. Developmental defects of the ventromedial hypothalamic nucleus and pituitary gonadotroph in the Ftz-F1 disrupted mice. Dev Dyn. 1995;204:22–9.
12. Almeida MQ, Soares IC, Ribeiro TC, Fragoso MC, Marins LV, Wakamatsu A, Ressio RA, Nishi MY, Jorge AA, Lerario AM, Alves VA, Mendonca BB, Latronico AC. Steroidogenic factor 1 overexpression and gene amplification are more frequent in adrenocortical tumors from children than from adults. J Clin Endocrinol Metab. 2010;95:1458–62.
13. Doghman M, Karpova T, Rodrigues GA, Arhatte M, De Moura J, Cavalli LR, Virolle V, Barbry P, Zambetti GP, Figueiredo BC, Heckert LL, Lalli E. Increased steroidogenic factor-1 dosage triggers adrenocortical cell proliferation and cancer. Mol Endocrinol. 2007;21:2968–87.
14. Sbiera S, Schmull S, Assie G, Voelker HU, Kraus L, Beyer M, Ragazzon B, Beuschlein F, Willenberg HS, Hahner S, Saeger W, Bertherat J, Allolio B, Fassnacht M. High diagnostic and prognostic value of steroidogenic factor-1 expression in adrenal tumors. J Clin Endocrinol Metab. 2010;95:E161–71.
15. Zubair M, Oka S, Parker KL, Morohashi K. Transgenic expression of Ad4BP/SF-1 in fetal adrenal progenitor cells leads to ectopic adrenal formation. Mol Endocrinol. 2009;23:1657–67.
16. Rabinowitz JD, Kimball E. Acidic acetonitrile for cellular metabolome extraction from Escherichia coli. Anal Chem. 2007;79:6167–73.

17. Nathan C, Ding A. SnapShot: reactive oxygen intermediates (ROI). Cell. 2010;140:e952.
18. Pollak N, Dolle C, Ziegler M. The power to reduce: pyridine nucleotides—small molecules with a multitude of functions. Biochem J. 2007;402:205–18.
19. Morohashi K, Baba T, Tanaka M. Steroid hormones and the development of reproductive organs. Sex Dev. 2013;7:61–79.
20. DeBerardinis RJ, Mancuso A, Daikhin E, Nissim I, Yudkoff M, Wehrli S, Thompson CB. Beyond aerobic glycolysis: transformed cells can engage in glutamine metabolism that exceeds the requirement for protein and nucleotide synthesis. Proc Natl Acad Sci U S A. 2007;104:19345–50.
21. Labuschagne CF, JF N, vd B, Mackay GM, Vousden KH, Maddocks OD. Serine, but not glycine, supports one-carbon metabolism and proliferation of cancer cells. Cell Rep. 2014;7:1248–58.
22. Lewis CA, Parker SJ, Fiske BP, McCloskey D, Gui DY, Green CR, Vokes NI, Feist AM, Vander Heiden MG, Metallo CM. Tracing compartmentalized NADPH metabolism in the cytosol and mitochondria of mammalian cells. Mol Cell. 2014;55:253–63.
23. Morohashi K. Gonadal and extragonadal functions of Ad4BP/SF-1: developmental aspects. Trends Endocrinol Metab. 1999;10:169–73.
24. Parker KL, Rice DA, Lala DS, Ikeda Y, Luo X, Wong M, Bakke M, Zhao L, Frigeri C, Hanley NA, Stallings N, Schimmer BP. Steroidogenic factor 1: an essential mediator of endocrine development. Recent Prog Horm Res. 2002;57:19–36.
25. Takayama K, Morohashi K, Honda S, Hara N, Omura T. Contribution of Ad4BP, a steroidogenic cell-specific transcription factor, to regulation of the human CYP11A and bovine CYP11B genes through their distal promoters. J Biochem. 1994;116:193–203.
26. Baba T, Otake H, Sato T, Miyabayashi K, Shishido Y, Wang CY, Shima Y, Kimura H, Yagi M, Ishihara Y, Hino S, Ogawa H, Nakao M, Yamazaki T, Kang D, Ohkawa Y, Suyama M, Chung BC, Morohashi K. Glycolytic genes are targets of the nuclear receptor Ad4BP/SF-1. Nat Commun. 2014;5:3634.
27. Inoue M, Shima Y, Miyabayashi K, Tokunaga K, Sato T, Baba T, Ohkawa Y, Akiyama H, Suyama M, Morohashi K. Isolation and characterization of fetal Leydig progenitor cells of male mice. Endocrinology. 2016;157:1222–33.
28. Shima Y, Miyabayashi K, Haraguchi S, Arakawa T, Otake H, Baba T, Matsuzaki S, Shishido Y, Akiyama H, Tachibana T, Tsutsui K, Morohashi K. Contribution of Leydig and Sertoli cells to testosterone production in mouse fetal testes. Mol Endocrinol. 2013;27:63–73.
29. Shima Y, Matsuzaki S, Miyabayashi K, Otake H, Baba T, Kato S, Huhtaniemi I, Morohashi K. Fetal Leydig cells persist as an androgen-independent subpopulation in the postnatal testis. Mol Endocrinol. 2015;29:1581–93.
30. Li B, Baba T, Miyabayashi K, Sato T, Shima Y, Ichinose T, Miura D, Ohkawa Y, Suyama M, Morohashi KI. Role of Ad4-binding protein/steroidogenic factor 1 in regulating NADPH production in adrenocortical Y-1 cells. Endocr J. 2017;64:315–24.
31. Freeman DA, Ascoli M. Studies on the source of cholesterol used for steroid biosynthesis in cultured Leydig tumor cells. J Biol Chem. 1982;257:14231–8.
32. Miller WL. Molecular biology of steroid hormone synthesis. Endocr Rev. 1988;9:295–318.
33. Fayard E, Auwerx J, Schoonjans K. LRH-1: an orphan nuclear receptor involved in development, metabolism and steroidogenesis. Trends Cell Biol. 2004;14:250–60.
34. Wilson TE, Fahrner TJ, Johnston M, Milbrandt J. Identification of the DNA binding site for NGFI-B by genetic selection in yeast. Science. 1991;252:1296–300.
35. Charest-Marcotte A, Dufour CR, Wilson BJ, Tremblay AM, Eichner LJ, Arlow DH, Mootha VK, Giguere V. The homeobox protein Prox1 is a negative modulator of ERR{alpha}/PGC-1{alpha} bioenergetic functions. Genes Dev. 2010;24:537–42.
36. Chaveroux C, Eichner LJ, Dufour CR, Shatnawi A, Khoutorsky A, Bourque G, Sonenberg N, Giguere V. Molecular and genetic crosstalks between mTOR and ERRalpha are key determinants of rapamycin-induced nonalcoholic fatty liver. Cell Metab. 2013;17:586–98.

4 Role of Lgr5-Expressing Stem Cells in Epithelial Renewal and Cancer in the Reproductive Tract

Annie Ng and Nick Barker

Abstract

The identification and molecular characterisation of somatic stem cells are pivotal to understanding the mechanisms of tissue development and homeostasis, as well as oncogenesis. The epithelia lining the female reproductive tract undergo successive cycles of cellular regeneration and remodelling. Recent lineage tracing experiments have formally shown that the leucine-rich repeat-containing G protein-coupled receptor 5 (Lgr5) marks OSE-resident stem cells that drive lifelong epithelial homeostasis and ovulatory wound repair in vivo. Deciphering Lgr5 stem-driven tissue maintenance may provide insights into the aetiology of human ovarian epithelial cancers.

Keywords

Lgr5 · Wnt · Ovary epithelium · Ovarian cancer · Stem cells · Self-renewal Regeneration · Wound repair

4.1 Introduction

In adult epithelia, somatic stem cells represent powerhouses that continuously supply new cells to replace those lost or damaged and thereby maintain optimal tissue function. Located within specialised niches, small populations of stem cells are

A. Ng
Centre for Innate Immunity and Infectious Diseases,
Hudson Institute of Medical Research, Clayton, VIC, Australia

N. Barker (✉)
A*STAR Institute of Medical Biology, Singapore, Singapore

Centre for Regenerative Medicine, The University of Edinburgh, Edinburgh, UK

Cancer Research Institute, Kanazawa University, Japan
e-mail: nicholas.barker@imb.a-star.edu.sg

H. Katabuchi et al. (eds.), *Cell Biology of the Ovary*,
https://doi.org/10.1007/978-981-10-7941-2_4

distinguished from other cell types by their ability to self-renew, while concomitantly generating at least one type of mature functional progeny to ensure tissue maintenance [1].

In mammals, the female reproductive tract is compartmentalised into the ovaries, fallopian tubes (oviducts in non-humans) and uterus. Each compartment is lined by epithelial cells that possess major regenerative capabilities. During ovulation, the ovarian surface epithelium (OSE) ruptures to facilitate the release of a mature oocyte, which is subsequently guided into the fallopian tube via its distal fimbrial fringes for fertilisation. The ovulatory epithelial stigma wound created on the ovary surface post-oocyte release is rapidly healed via increased proliferation and migration at wound margins, while damage sustained by the delicate fimbrial epithelial fringes as a result of exposure to the inflammatory, reactive oxygen species-rich ovulatory follicular fluid [2, 3] is repaired by cellular replenishment. The endometrium and endocervix also engage in dynamic proliferation and shedding associated with ovulation and pregnancy. The enormous capacity of the reproductive epithelia to self-renew and undergo profound cellular remodelling during hundreds of ovulatory cycles likely reflects the existence of stem cell-driven regenerative mechanisms [4]. However, identifying and characterising epithelial stem cells in the female reproductive tract have proven challenging due to the absence of validated stem cell markers that can be used for their prospective isolation and functional study.

Somatic stem cells that express *Lgr5* have been documented in a variety of epithelial tissues, including the intestinal tract [5], stomach [6], kidney [7], mammary gland [8] and hair follicle [9]. *Lgr5* is a target of the canonical Wnt signalling which is well recognised for its role in development and adult stem cell maintenance, as well as in tumour initiation and growth [1, 6, 10, 11]. Activation of Wnt signalling relies on the binding of Wnt ligands to their cognate Frizzled and LRP protein receptors to stabilise and release β-catenin from the adenomatous polyposis coli (APC)-axis inhibition protein (AXIN) destruction complex; β-catenin then translocates into the nucleus where it interacts with lymphoid enhancer-binding factor (LEF)/T-cell factor (TCF) transcription factors to activate cell type-specific target genes, including the expression of *Lgr5* [12]. *Lgr5* is also a receptor for stem cell factor R-spondins (Rspo1–4), binding of which strongly potentiates local Wnt signals within $Lgr5^+$ stem cells [13].

Expression analyses using single-molecule fluorescent in situ hybridisation, quantitative polymerase chain reaction and *Lgr5-EGFP* reporter models (i.e. using the *Lgr5-EGFP-ires-CreER*^{T2} knock-in mice, in which enhanced green fluorescent protein is driven by the *Lgr5* promoter) have documented $Lgr5^+$ epithelial cell populations throughout the female reproductive tract encompassing the ovary [14, 15], fallopian tube/oviduct fimbria [14] and uterus [16, 17] including the cervix [18]. In the ovary, $Lgr5^+$ OSE cells represent bona fide stem cells that contribute to lifelong epithelial homeostasis and post-ovulation wound repair in vivo [14, 15]. During development, *Lgr5* marks stem cell precursors that give rise to the adult OSE and cortical granulosa cell ($Lgr5^-Foxl2^+$) lineages [14, 19], as well as the adult oviduct/fimbrial epithelial cell lineage [14]. In humans, *LGR5* is expressed in the OSE and fallopian tube fimbrial epithelium [14]. The identification of $LGR5^+$ cells in the

OSE and fimbrial epithelium in humans may have major implications, as both cell types are believed to represent potential origins of ovarian cancer.

In this chapter, we provide an overview on the methodologies employed thus far to identify epithelial stem cells in the female reproductive tract, with an emphasis on the discovery of stem cells expressing *Lgr5* within the OSE and ovary-fimbria junctional epithelium. We also explore future directions that can improve our understanding of stem cell biology in this dynamic tissue and potentially decipher any role that $Lgr5^+$ stem cells may play as cells of origin in human epithelial ovarian cancer.

4.2 Methods Employed to Identify and Characterise Reproductive Epithelial Stem Cells

The lack of established biomarkers has led researchers to employ a variety of retrospective methodologies to identify cells with functional stem cell properties in the reproductive tract epithelia.

A common means to identify stem cell candidates is the label-retaining pulse-chase assay, which exploits the fact that stem cells are traditionally thought to divide infrequently. All cells in the tissue are first 'pulsed' for a defined time with the DNA nucleoside 5-bromodeoxyuridine (BrdU) and/or genetically manipulated to express an inducible histone-linked green fluorescent protein (*H2B-GFP* transgenic mice). Following DNA label removal and a prolonged 'chase' period to facilitate cell division, highly proliferative cells dilute out the chromatin label to undetectable levels, while slow-cycling stem cells retain their label and can be visualised in situ as label-retaining cells. Using this surrogate stemness approach, epithelial label-retaining cells have been described on the ovary surface [20], ovary hilum [15], fimbria [21, 22], uterus [23] and endocervix [22]. Functional in vitro assays further highlighted the stem cell characteristics of purified label-retaining cells, notably their ability to form self-renewing colonies or spheres that contained all mature lineages of the native epithelium. In humans, three stem-like epithelial populations with long-term clonal expansion and differentiation properties have also been localised to the fallopian tube fimbria, which lends support to the notion that repeated exposure of the most distal portion of the fallopian tube to the harsh ovulatory environment imposes a requirement for stem-driven regenerative repair mechanisms. While it remains unclear whether these populations represent distinct stem cell pools, they are phenotypically defined by the expression of *CD44* and *α6-integrin* [24]; *LGR6, AXIN2* and *Olfactomedin4* [25]; or *LGR6* and *LRIG1* [26].

In parallel, flow cytometry approaches have been used to isolate stem cell-enriched 'side populations' in the OSE [27] and endometrium [28] that effectively efflux the DNA-binding dye Hoechst 33342 through expression of ATP-binding cassette transporters, as well as an OSE population demonstrating elevated activity of the aldehyde dehydrogenase (Aldh) family of detoxifying enzymes [15]. An enhanced ability to remove foreign molecules from their cytoplasm and self-protect against drug toxicity are presumptive stem cell characteristics. An isolated $Aldh^{hi}$

OSE fraction displayed self-renewal properties ex vivo and co-expressed *Lgr5* at the ovary-fimbria junctional zone, thus establishing an $Aldh^{hi}Lgr5^{+}$ stem cell niche at the ovary hilum [15] (Fig. 4.1). In addition to the ovary hilum, clusters of $Lgr5^{+}$ OSE cells have also been documented at cleft regions between growing follicles and in the vicinity of ovulatory wound stigma sites [14] (Fig. 4.1).

Recently, the stem cell potential of $Lgr5^{+}$ ($Aldh^{hi}$) OSE cells at the ovary hilum [15] and throughout the ovary surface [14] was evaluated via in vivo genetic lineage tracing. Lineage tracing involves the introduction of a heritable reporter gene into candidate stem cells so that their fate and the fate of all of their progeny can be tracked over time within their natural environment (Fig. 4.2). This technique has been successfully used to formally establish $Lgr5^{+}$ stem cell identity in the intestinal tract [5], stomach [6], mammary gland [8] and hair follicle [9]. To trace $Lgr5^{+}$ ovary populations, *Lgr5-EGFP-ires-CreER*^{T2} mice were first crossed with a conditional lineage reporter that harboured either a histochemical (*Rosa26-loxP-stop-loxP-LacZ*) (Fig. 4.2) or fluorescent (*Rosa26-loxP-stop-loxP-tdTomato*, *Rosa26-loxP-stop-loxP-EYFP* or *Rosa26-loxP-stop-loxP-4colour*) lineage reporter. Administration of a single dose of tamoxifen led to Cre-induced removal of the intervening 'stop' sequence and activation of the lineage reporter in $Lgr5^{+}$ cells and their progeny (Fig. 4.2). Comprehensive tracing analyses formally established hilum and extrahilum $Lgr5^{+}$ populations to be long-lived stem cells that contribute to the normal homeostasis of the ovary epithelium during the reproductive lifetime

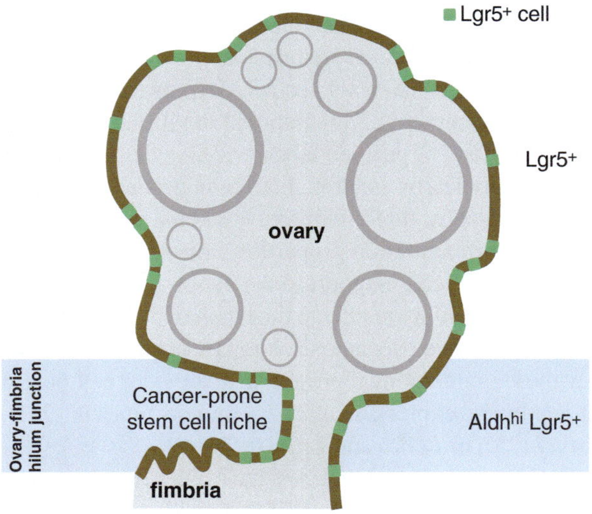

Fig. 4.1 Schematic showing mouse *Lgr5-EGFP* expression (*green cells*) in OSE at inter-follicular cleft regions and ovary-fimbria hilum junction (*shaded area*). The ovary-fimbria junctional epithelia display high activity of the Aldh family of detoxifying enzymes and are a stem cell-enriched region. Mouse studies have revealed the propensity of the ovary-fimbria junctional epithelia to neoplastic transformation in vivo

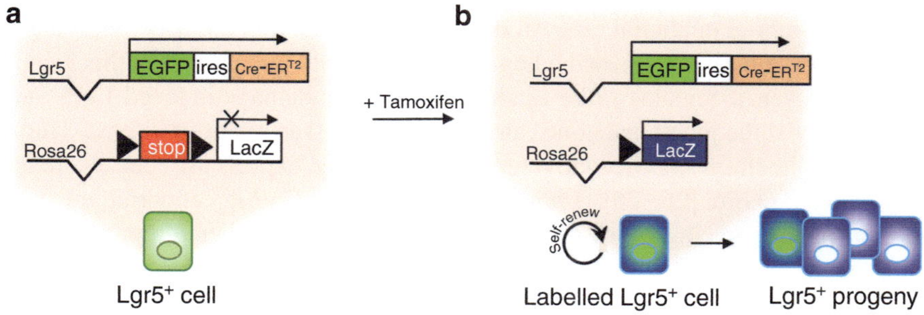

Fig. 4.2 Lgr5$^+$ cell-driven in vivo lineage tracing strategy. (**a**) Mice are generated which harbour an inducible *Lgr5* promoter-driven *CreER^{T2}* activator and a conditional *Rosa26* promoter-driven lineage reporter, such as *LacZ* which encodes the β-galactosidase gene. An upstream loxP (*triangle*)-flanked 'stop' sequence prevents the transcription of *LacZ*. (**b**) Upon tamoxifen administration, activated Cre removes the loxP-flanked 'stop' sequence to enable transcription of *LacZ* in Lgr5-EGFP$^+$ cells. Labelled Lgr5-EGFP$^+$ cells pass on the *LacZ* lineage mark to their progeny, which can be visualised in situ by a blue cytoplasmic precipitate after β-galactosidase staining

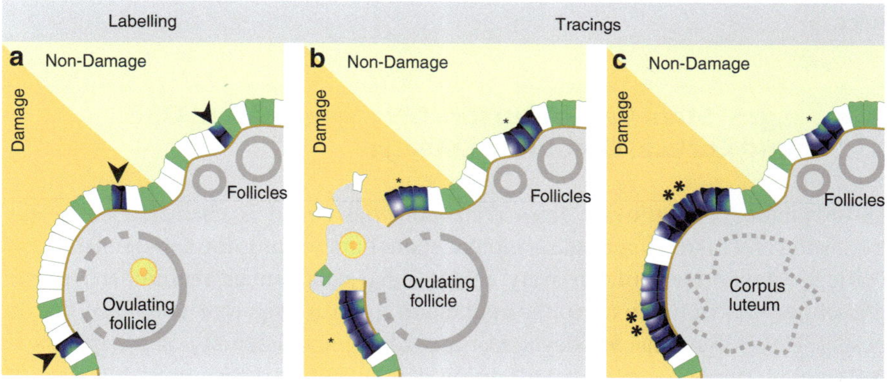

Fig. 4.3 Cartoon depicting Lgr5$^+$ OSE lineage tracings. (**a**) Administration of a single dose of tamoxifen labels a subset of Lgr5$^+$ OSE stem cells (*green + blue cells*, highlighted by *arrowheads*). (**b**) Over time, labelled Lgr5$^+$ OSE stem cells generate LacZ progeny (highlighted by *) that are Lgr5$^+$ (*green + blue cells*) and Lgr5$^-$ (*blue cells*) throughout the ovary surface. (**c**) Lgr5$^+$ OSE cell output is highly variable, with the largest 'lineage clones' invariably located at wound margins of ovulatory follicles and corpus lutea ('damage', highlighted by **) compared to follicular-growth regions ('non-damage', highlighted by *), consistent with a role of Lgr5$^+$ OSE stem cells in ovulatory epithelial regenerative repair in vivo

of a mouse (16 months) [14, 15] (Fig. 4.3). Strikingly, the output of individual Lgr5$^+$ cells on the ovary surface was highly variable, with the largest 'lineage clones' invariably located at ovulatory wound damage regions and atop corpora lutea (post-ovulatory follicles) [14] (Fig. 4.3). This is consistent with increased proliferative activity of ovulation site adjacent Lgr5$^+$ OSE cells [14] and the notion that post-ovulatory repair enforces the highest rate of cellular replenishment to achieve rapid restoration of the ovary epithelial barrier.

In contrast, $Lgr5^+$ cells residing at the ovary-fimbria junctional zone do not contribute to fimbrial epithelial renewal during steady-state homeostasis, implying the existence of non-*Lgr5*-expressing homeostatic stem cell populations at this location [14]. However, it remains to be established whether damage-induced $Lgr5^+$ stem cell activity at the junctional zone can contribute to tissue repair or disease. Human fimbrial epithelium was recently reported to highly express the closely related homologue, *LGR6* [25, 26]. Using *Lgr6-EGFP-ires-CreER*^{T2} reporter mice and in vivo lineage tracing, *Lgr6* has been established as a marker of epithelial stem cells in the epidermis [29]. The same mouse model would be equally useful in evaluating the contribution of $Lgr6^+$ fimbrial cells to in vivo tissue homeostasis and repair in the mouse oviduct.

Endogenous *Lgr5* expression has also been detected in the glandular and luminal epithelia lining the uterus [16, 17], as well as cervical epithelium [18]. The stem cell potential of $Lgr5^+$ cells in these regenerative epithelial compartments has not been reported in the *Lgr5-EGFP-ires-CreER*^{T2} reporter mice [14, 15] due to silencing of the transgene within these tissues. Future work awaits the generation of more robust knock-in mouse models to examine the potential involvement of these $Lgr5^+$ epithelial populations in stem-driven remodelling throughout the rest of the reproductive tract.

4.3 $Lgr5^+$ Stem Cell Precursors Give Rise to Adult OSE and Cortical Granulosa Lineages

Differentiation of the ovary from the bipotential gonad is characterised by the expansion and ingression of the coelomic epithelium towards the sub-epithelial and cortical compartments of the ovary anlage and a concomitant recruitment of pre-granulosa cells from the coelomic epithelium to form ovigerous cords and oocyte nests [30, 31]. Subsequent oocyte nest breakdown and assembly of pre-granulosa cells around single oocytes form the primordial follicles that support lifelong female fertility.

Ovarian differentiation is tightly controlled by two *Wnt* factors, *secreted wingless-type MMTV integration site family 4* (*Wnt4*) and *Rspo1* [32]. Interestingly, the coelomic epithelium and the cortical region of the mouse ovary anlage harbour *Lgr5*-marked stem cell precursors whose expression is under *Rspo1-Wnt4* regulation [14, 19]. Developmental in vivo lineage tracings have revealed that these $Lgr5^+$ stem cell precursors contributed to the future OSE ($Foxl2^-$) [14] and granulosa cells ($Foxl2^+$) of adult cortical follicles [14, 19] through a $Lgr5^+Foxl2^+$ progenitor/phenotypic intermediary [19, 33]. A role for *a disintegrin and metalloproteinase domain 10 (Adam10)* via the Notch signalling pathway has been proposed in Lgr5-driven granulosa cell specification, as treatment of cultured perinatal mouse ovaries with Adam10-Notch inhibitors leads to reduced proliferation of $Lgr5^+$ ovarian precursors coupled with blockage of their differentiation into $Foxl2^+$ granulosa cells [34]. Adam10-Notch regulates many $Lgr5^+$ stem-driven epithelia, impacting on the survival and fate decisions between secretory and absorptive epithelial lineages of

Lgr5$^+$ crypt base columnar stem cells in the intestinal tract [35, 36], as well as proliferation and differentiation of all mature epithelia of Lgr5$^+$ gastric stem cells [37].

Lgr5 is also highly expressed in the anterior Müllerian duct, the primordia of the adult oviduct including fimbria. These Müllerian Lgr5$^+$ precursors contributed to the epithelial cell lineages of the oviduct including fimbria, as well as the ovary-fimbria junctional epithelium in vivo [14].

4.4 Regulation of Lgr5$^+$ Stem Cell-Driven OSE Regeneration: Intrinsic Versus Niche Signals

Stem cell behaviour is determined by cell-autonomous programmes and instructive paracrine signals derived from the stem cell niches. While the molecular mechanisms that govern Lgr5$^+$ OSE stem cell fate remain to be defined, candidate signal regulators include *Wnt* stemness-promoting and *Bmp/Tgf-β* differentiation-promoting signals. *Adam10-Notch* signals may also continue to operate in adulthood by maintaining dormancy of Lgr5$^+$ OSE stem cells during non-damage homeostasis.

The canonical Wnt pathway is recognised for its role in stem cell maintenance and fate determination [1, 11]. Expression profiling has revealed that Lgr5-EGFP^{hi} OSE stem cells are enriched in expression for *Wnt4* and the tumour necrosis factor receptor family member *Troy*, a Wnt target gene [14]. The ability to produce their own *Wnt4* ligands may facilitate autocrine Wnt signalling within Lgr5$^+$ OSE stem cells to confer stemness and enhance survival. *Troy*, a known inhibitor of Wnt signalling in Lgr5$^+$ intestinal and gastric stem cells [38, 39], may fine-tune internal Wnt signal strength to prevent over-proliferation of OSE stem cells. Presumptive Lgr5$^+$ OSE niches encompass the underlying ovulatory follicle and its follicular fluid, with Lgr5$^+$ OSE stem cells closest to rupture wound margins subjected to the strongest paracrine effects from the follicular milieu compared to wound-distant Lgr5$^+$ cell counterparts. Ovulatory follicular fluid contains ovarian hormones (oestrogen and progesterone) and circulating gonadotropins, as well as a myriad of OSE mitogenic growth factors, notably from the EGF [40–42], TGF-β [43, 44], HGF [45, 46] and KGF [45, 47] families. Surface stroma cells bordering the OSE represent additional niche candidates as they express high levels of all TGF-β ligands [48, 49]. Whether Lgr5$^+$ OSE stem cells express their associated receptors remains to be determined.

Of note, a role for *Bmp* signals in regulating Lgr5$^+$ OSE stem cells has been recently explored. Treatment of cultured human foetal ovary epithelial cells with BMP4 led to a marked up-regulation of *LGR5* expression [50], thus demonstrating that LGR5 is a transcriptional target of BMP4 signalling in human foetal ovary. The *TGF-β/BMP* signalling pathway is known for its involvement in wound healing by promoting EMT in many epithelia, including OSE cells [51]. BMP4 induces primary human ovarian cancer cells to switch from the typical cobblestone to a flattened morphology resembling cell spreading with enhanced adherence [52]. In rat, in situ hybridization studies have revealed that *Bmp4* is expressed in OSE cells atop ovulating follicles and their post-ovulatory corpora lutea [53]. Such a spatiotemporal *Bmp4* expression adjacent to ovulatory follicles may create an EMT-promoting gradient to achieve highly controlled wound repair. It is plausible that Bmp4 signals at an ovulatory follicular apex representing a future stigma site stimulate $Lgr5^+$ OSE stem cells located at wound margins to differentiate and activate EMT for efficient stigma repair. Conversely, absence of Bmp4 signals at wound-distant niche would maintain $Lgr5^+$ OSE stem cells in a dormant state.

Future experiments designed to reveal the composition and functional impact of autocrine and paracrine signalling on $Lgr5^+$ OSE stem cells will be invaluable in understanding how these cells regulate stemness and fate decisions in response to local damage signals. Definitive characterization of any OSE and fimbrial stem cell niche awaits in vivo studies targeting modulation of candidate growth factors and signalling pathways in the ovary or the development of ex vivo stem cell-driven OSE culture systems using defined growth media.

4.5 Cancer of the Reproductive Tract: $Lgr5^+$ OSE Stem Cells as Ovarian Cancer Origins

Cancers of the reproductive tract are classified as ovarian, uterine or cervical depending on the tissue of origin. Some ovarian cancers, notably the most common and deadliest high-grade serous ovarian carcinomas (HGSC) subtype, may arise from the fallopian tube fimbria. The first evidence of an extra-ovarian origin for ovarian epithelial cancers came from the discovery of preinvasive secretory cell outgrowths (SCOUTS) and/or serous tubal intraepithelial carcinomas (STIC) lesions in the fallopian tubes of women carrying hereditary *breast cancer susceptibility gene* (*BRCA*) mutations [54], or p53 mutation signatures—believed to be SCOUT/STIC precursors—in histologically benign fallopian tubes of women, regardless of their *BRCA* mutation status [54–56].

The study of ovarian cancer mouse models has indicated that both the ovary [15, 57–64] and fallopian tube [63, 65–69] are capable of neoplastic transformation. However, the culprit cellular origin of each ovarian cancer subtype remains unknown. Any cell type can acquire oncogenic mutations to become neoplastic, but stem cells, given their longevity due to intrinsic self-renewal, are at greater risk of accumulating the necessary multiple mutational load required to initiate cancer formation.

OSE-resident $Lgr5^+$ stem cells on the ovary surface and at the ovary-fimbria junctional zone at the ovary hilum therefore represent potential cancer-initiating cells. Indeed, $Lgr5^+$ stem cells in the gastrointestinal tract have previously been shown to serve as key cancer origins following mutation [6, 10]. Conditional targeting of key ovarian oncogenes and tumour suppressors within $Lgr5^+$ stem populations using *Lgr5-EGFP-ires-CreER*^{T2} mouse models should directly address their contribution to tumour initiation and/or malignancy potential in vivo. Recently, SCOUT/STIC precursor lesions were observed to harbour hyperactive WNT signalling, as evidenced by strong nuclear accumulation of β-catenin and elevated LEF1 [70]. Deregulated WNT signalling may therefore constitute an early event in HGSC carcinogenesis. A role for activated Wnt in stem-driven carcinogenesis is well-validated for other tissues. Constitutive Wnt activation, via deletion of *Apc*, in $Lgr5^+$ crypt base columnar stem cells and $Lgr5^+$ gastric stem cells gives rise to rapid development of self-renewing adenomas in the intestinal tract [10] and stomach [6], respectively. Importantly, deregulation of Wnt signalling in intestinal transit-amplifying and differentiated epithelia fails to drive cancer formation in vivo [10], thus identifying intestinal stem cells as a key target of Wnt-driven cancer in the gastrointestinal tract.

It is important to note that acquisition of similar mutation spectra by different cell types in a tissue leads to the generation of distinct cancer subtypes. This was recently illustrated in a comparative study of in vivo tumours originating from the OSE and oviductal epithelium following inactivation of *Apc* and *Pten* [63]. Using $Ovgp1\text{-}iCreER^{T2};Apc^{fl/fl};Pten^{fl/fl}$ transgenic mice (which harbour an inducible *iCre* driven by the oviductal secretory epithelium-specific *Ovgp1* promoter), conditional deletion of *Apc* and *Pten* was achieved in the oviductal epithelium by intraperitoneal administration of tamoxifen and in the OSE by direct *iCre* delivery using the adenovirus (*AdCre*) into the ovarian bursa space. Although tumours developed from both the OSE and fimbria, lending support to the notion that both cell types can initiate ovarian carcinogenesis, the tumours differed markedly in their histopathology and molecular signatures, particularly in *Wnt* target gene expression. Of note, *Lgr5* expression was present and up-regulated in the oviductal, but not OSE, tumours [63], consistent with an expansion of the $Lgr5^+$ stem cell pool during the transformation process. As the molecular signature of oviduct tumours more closely resembled that of human ovarian cancers, this has led to the conclusion that transformation of an oviductal epithelial cell (instead of OSE) is likely to be the initiating event in ovarian carcinogenesis. However, multiple cells of origins likely account for the intrinsic heterogeneity of the ovarian epithelial cancer subtypes.

The $Lgr5^+Aldh^{hi}$ ovary-fimbria junctional zone has been intensively investigated as a potential cancer source. Epithelia of the ovary-fimbria junction appear more susceptible to targeted transformation, as evidenced by the development of atypical lesions in the ovary hilum but not other parts of the ovary following targeted inactivation of conditional *Trp53* and *Rb1* alleles in OSE by intrabursal *AdCre* administration [15]. Moreover, only micro-dissected *Trp53;Rb1*-deficient OSE cells of the ovary hilum induced HGSC-like tumours in vivo, with evidence of lung metastasis in a subset of the transplanted mouse cohort [15]. Whether these tumorigenic hilum OSE cells expressed *Lgr5* was not determined, but these cells were high in *Aldh1* expression and therefore presumably also *Lgr5*. Collectively, these key findings define an epithelial cell within the ovary-fimbria junction as a potential source of ovarian cancer.

The ovary-fimbria continuum is conserved in humans. Termed ovarian fimbriae [71], the human ovary-fimbria junctional zone represents a microanatomic isthmus lined with 'transitional' epithelia that share OSE and fimbrial epithelial gene expression [72] and are enriched in the expression of several stem cell genes including *ALDH1A1/1A2*, *NANOG*, *LHX9* and *SFRP1* [71]. Whether *LGR5* is also expressed in the ovarian fimbriae remains to be determined. However, highly sensitive single-molecule fluorescent in situ hybridisation combined with qPCR has revealed *LGR5* expression in the fimbrial epithelial fringes [14], although the presence of *LGR5* has not been confirmed in human fimbrial epithelium-derived cell lines and 3D organoids [25, 26]. The reason for this discrepancy is unclear, but $LGR5^+$ fimbrial epithelia may not be amendable to sustained growth in available culture systems, or *LGR5* expression may be confined to a small subset of fallopian tube epithelia at the fimbrial fringe tips.

The concept of a heightened cancer prevalence within the junctional zone bordering two epithelial cell types has also been observed at the esophago-gastric, corneal-limbus and endo-ectocervix regions [73]. In cervical cancers associated with the high-risk human papillomavirus, the appearance of premalignant cervical intraepithelial neoplasia (CIN) precursors invariably occurred within the squamocolumnar junction [74], where the squamous epithelium of the ectocervix abuts the columnar epithelium of the endocervix [75]. It has been purported that a reserve or basal cell residing within the squamocolumnar junction is the initiating cell of neoplastic change [75, 76]. In support of high stem cell activity within this region, a c-kit^+p63^+ basal label-retaining cell pool [21] and a basal *Sox2*-expressing pool have been reported in the mouse endocervical region. Lineage tracing experiments have further revealed that these $Sox2^+$ basal cells serve as stem/progenitor cells of the cervical epithelium in vivo [77].

4.6 Future Directions and Conclusion

Three-dimensional (3D) organoid cultures have proven to be ideal ex vivo systems for evaluating stem cell identity and studying cell intrinsic and extrinsic regulation of stem cell behaviour in both mice and humans [6, 78]. For less accessible tissues delivering limited quantities of experimental material such as the ovary and fallopian tube, development of near-physiological 3D organoids facilitating the rapid expansion and storage of epithelia for stem niche studies is of vital importance. Recently, 3D epithelial organoids that can be maintained long term (>16 months) have been successfully established from the human fallopian tube [25]. Robust growth relied on WNT signals (WNT3a and RSPO1) and EGF, while inhibition of TGF-β signalling was crucial for long-term expansion. These fallopian tube organoids expressed several stem cell markers including *hTERT*, *Olfactomedin-4* and *LGR6* but not *LGR5* [25]. Future studies should employ similar culture conditions to establish OSE-derived 3D organoids. This ex vivo system would be particularly amenable to performing real-time clonal tracking of Lgr5-$EGFP^+$ OSE stem cell behaviour as a means of better understanding the contribution of $Lgr5^+$ stem cells to epithelial maintenance and repair.

Deciphering fate decisions of $Lgr5^+$ (and other) stem cells in the ovary during homeostasis and disease and probing the contribution of the niche microenvironment to cancer initiation are important aspects to address in the ovarian cancer field. Notably, investigation of the tumour initiation potential of $Lgr5^+Aldh1^{hi}$ stem cells within the ovary-fimbria junctional zone following genetic alteration of key ovarian genes requires further investigation. Junctional Lgr5$^+$ epithelia may be subject to overlapping paracrine signals from the two distinct microenvironments of the ovary and fimbria. Though speculative, the pro-inflammatory and mitogenic factor-rich ovarian microenvironment may provide a particularly damaging environment that endows dislodged Lgr5$^+$ stem cells with uncontrolled proliferation, dysregulated self-renewal and tumour initiation.

References

1. Reya T, Morrison SJ, Clarke MF, Weissman IL. Stem cells, cancer, and cancer stem cells. Nature. 2001;414:105–11.
2. Bahar-Shany K, Brand H, Sapoznik S, Jacob-Hirsch J, Yung Y, Korach J, Perri T, Cohen Y, Hourvitz A, Levanon K. Exposure of fallopian tube epithelium to follicular fluid mimics carcinogenic changes in precursor lesions of serous papillary carcinoma. Gynecol Oncol. 2014;132:322–7.
3. Huang HS, Chu SC, Hsu CF, Chen PC, Ding DC, Chang MY, Chu TY. Mutagenic, surviving and tumorigenic effects of follicular fluid in the context of p53 loss: initiation of fimbria carcinogenesis. Carcinogenesis. 2015;36:1419–28.
4. Ng A, Barker N. Ovary and fimbrial stem cells: biology, niche and cancer origins. Nat Rev Mol Cell Biol. 2015;16:625–38.
5. Barker N, van Es JH, Kuipers J, Kujala P, van den Born M, Cozijnsen M, Haegebarth A, Korving J, Begthel H, Peters PJ, Clevers H. Identification of stem cells in small intestine and colon by marker gene Lgr5. Nature. 2007;449:1003–7.
6. Barker N, Huch M, Kujala P, van de Wetering M, Snippert HJ, van Es JH, Sato T, Stange DE, Begthel H, van den Born M, Danenberg E, van den Brink S, Korving J, Abo A, Peters PJ, Wright N, Poulsom R, Clevers H. Lgr5(+ve) stem cells drive self-renewal in the stomach and build long-lived gastric units in vitro. Cell Stem Cell. 2010;6:25–36.
7. Barker N, Rookmaaker MB, Kujala P, Ng A, Leushacke M, Snippert H, van de Wetering M, Tan S, Van EJH, Huch M, Poulsom R, Verhaar MC, Peters PJ, Clevers H. Lgr5(+ve) stem/progenitor cells contribute to nephron formation during kidney development. Cell Rep. 2012;2:540–52.
8. NY F, Rios AC, Pal B, Law CW, Jamieson P, Liu R, Vaillant F, Jackling F, Liu KH, Smyth GK, Lindeman GJ, Ritchie ME, Visvader JE. Identification of quiescent and spatially restricted mammary stem cells that are hormone responsive. Nat Cell Biol. 2017;19:164–76.
9. Jaks V, Barker N, Kasper M, van Es JH, Snippert HJ, Clevers H, Toftgard R. Lgr5 marks cycling, yet long-lived, hair follicle stem cells. Nat Genet. 2008;40:1291–9.
10. Barker N, Ridgway RA, van Es JH, van de Wetering M, Begthel H, van den Born M, Danenberg E, Clarke AR, Sansom OJ, Clevers H. Crypt stem cells as the cells-of-origin of intestinal cancer. Nature. 2009;457:608–11.
11. Reya T, Clevers H. Wnt signalling in stem cells and cancer. Nature. 2005;434:843–50.
12. Van der Flier LG, Sabates-Bellver J, Oving I, Haegebarth A, De Palo M, Anti M, Van Gijn ME, Suijkerbuijk S, Van de Wetering M, Marra G, Clevers H. The intestinal Wnt/TCF signature. Gastroenterology. 2007;132:628–32.
13. de Lau W, Barker N, Low TY, Koo BK, Li VS, Teunissen H, Kujala P, Haegebarth A, Peters PJ, van de Wetering M, Stange DE, van Es JE, Guardavaccaro D, Schasfoort RB, Mohri Y, Nishimori K, Mohammed S, Heck AJ, Clevers H. Lgr5 homologues associate with Wnt receptors and mediate R-spondin signalling. Nature. 2011;476:293–7.
14. Ng A, Tan S, Singh G, Rizk P, Swathi Y, Tan TZ, Huang RY, Leushacke M, Barker N. Lgr5 marks stem/progenitor cells in ovary and tubal epithelia. Nat Cell Biol. 2014;16:745–57.
15. Flesken-Nikitin A, Hwang CI, Cheng CY, Michurina TV, Enikolopov G, Nikitin AY. Ovarian surface epithelium at the junction area contains a cancer-prone stem cell niche. Nature. 2013;495:241–5.

16. Sun X, Jackson L, Dey SK, Daikoku T. In pursuit of leucine-rich repeat-containing G protein-coupled receptor-5 regulation and function in the uterus. Endocrinology. 2009;150:5065–73.
17. Krusche CA, Kroll T, Beier HM, Classen-Linke I. Expression of leucine-rich repeat-containing G-protein-coupled receptors in the human cyclic endometrium. Fertil Steril. 2007;87:1428–37.
18. Sun B, Ye X, Li Y, Zhang W. Lgr5 is a potential prognostic marker in patients with cervical carcinoma. Int J Clin Exp Pathol. 2015;8:1783–9.
19. Rastetter RH, Bernard P, Palmer JS, Chassot AA, Chen H, Western PS, Ramsay RG, Chaboissier MC, Wilhelm D. Marker genes identify three somatic cell types in the fetal mouse ovary. Dev Biol. 2014;394:242–52.
20. Szotek PP, Chang HL, Brennand K, Fujino A, Pieretti-Vanmarcke R, Lo Celso C, Dombkowski D, Preffer F, Cohen KS, Teixeira J, Donahoe PK. Normal ovarian surface epithelial label-retaining cells exhibit stem/progenitor cell characteristics. Proc Natl Acad Sci U S A. 2008;105:12469–73.
21. Patterson AL, Pru JK. Long-term label retaining cells localize to distinct regions within the female reproductive epithelium. Cell Cycle. 2013;12:2888–98.
22. Wang Y, Sacchetti A, van Dijk MR, van der Zee M, van der Horst PH, Joosten R, Burger CW, Grootegoed JA, Blok LJ, Fodde R. Identification of quiescent, stem-like cells in the distal female reproductive tract. PLoS One. 2012;7:e40691.
23. Chan RW, Gargett CE. Identification of label-retaining cells in mouse endometrium. Stem Cells (Dayton, OH). 2006;24:1529–38.
24. Paik DY, Janzen DM, Schafenacker AM, Velasco VS, Shung MS, Cheng D, Huang J, Witte ON, Memarzadeh S. Stem-like epithelial cells are concentrated in the distal end of the fallopian tube: a site for injury and serous cancer initiation. Stem Cells. 2012;30:2487–97.
25. Kessler M, Hoffmann K, Brinkmann V, Thieck O, Jackisch S, Toelle B, Berger H, Mollenkopf HJ, Mangler M, Sehouli J, Fotopoulou C, Meyer TF. The Notch and Wnt pathways regulate stemness and differentiation in human fallopian tube organoids. Nat Commun. 2015;6:8989.
26. Yamamoto Y, Ning G, Howitt BE, Mehra K, Wu L, Wang X, Hong Y, Kern F, Wei TS, Zhang T, Nagarajan N, Basuli D, Torti S, Brewer M, Choolani M, McKeon F, Crum CP, Xian W. In vitro and in vivo correlates of physiological and neoplastic human Fallopian tube stem cells. J Pathol. 2016;238:519–30.
27. Gamwell LF, Collins O, Vanderhyden BC. The mouse ovarian surface epithelium contains a population of LY6A (SCA-1) expressing progenitor cells that are regulated by ovulation-associated factors. Biol Reprod. 2012;87:80.
28. Masuda H, Matsuzaki Y, Hiratsu E, Ono M, Nagashima T, Kajitani T, Arase T, Oda H, Uchida H, Asada H, Ito M, Yoshimura Y, Maruyama T, Okano H. Stem cell-like properties of the endometrial side population: implication in endometrial regeneration. PLoS One. 2010;5:e10387.
29. Snippert HJ, Haegebarth A, Kasper M, Jaks V, van Es JH, Barker N, van de Wetering M, van den Born M, Begthel H, Vries RG, Stange DE, Toftgard R, Clevers H. Lgr6 marks stem cells in the hair follicle that generate all cell lineages of the skin. Science. 2010;327:1385–9.
30. Mork L, Maatouk DM, McMahon JA, Guo JJ, Zhang P, McMahon AP, Capel B. Temporal differences in granulosa cell specification in the ovary reflect distinct follicle fates in mice. Biol Reprod. 2012;86:37.
31. Auersperg N, Wong AS, Choi KC, Kang SK, Leung PC. Ovarian surface epithelium: biology, endocrinology, and pathology. Endocr Rev. 2001;22:255–88.
32. Chassot AA, Bradford ST, Auguste A, Gregoire EP, Pailhoux E, de Rooij DG, Schedl A, Chaboissier MC. WNT4 and RSPO1 together are required for cell proliferation in the early mouse gonad. Development. 2012;139:4461–72.
33. Suzuki H, Kanai-Azuma M, Kanai Y. From sex determination to initial folliculogenesis in mammalian ovaries: morphogenetic waves along the anteroposterior and dorsoventral axes. Sex Dev. 2015;9:190–204.
34. Feng L, Wang Y, Cai H, Sun G, Niu W, Xin Q, Tang X, Zhang J, Wang C, Zhang H, Xia G. ADAM10-Notch signaling governs the recruitment of ovarian pregranulosa cells and controls folliculogenesis in mice. J Cell Sci. 2016.
35. VanDussen KL, Carulli AJ, Keeley TM, Patel SR, Puthoff BJ, Magness ST, Tran IT, Maillard I, Siebel C, Kolterud A, Grosse AS, Gumucio DL, Ernst SA, Tsai YH, Dempsey PJ, Samuelson

LC. Notch signaling modulates proliferation and differentiation of intestinal crypt base columnar stem cells. Development. 2012;139:488–97.
36. Tsai YH, VanDussen KL, Sawey ET, Wade AW, Kasper C, Rakshit S, Bhatt RG, Stoeck A, Maillard I, Crawford HC, Samuelson LC, Dempsey PJ. ADAM10 regulates Notch function in intestinal stem cells of mice. Gastroenterology. 2014;147:822–34. e813
37. Demitrack ES, Gifford GB, Keeley TM, Carulli AJ, VanDussen KL, Thomas D, Giordano TJ, Liu Z, Kopan R, Samuelson LC. Notch signaling regulates gastric antral LGR5 stem cell function. EMBO J. 2015;34:2522–36.
38. Fafilek B, Krausova M, Vojtechova M, Pospichalova V, Tumova L, Sloncova E, Huranova M, Stancikova J, Hlavata A, Svec J, Sedlacek R, Luksan O, Oliverius M, Voska L, Jirsa M, Paces J, Kolar M, Krivjanska M, Klimesova K, Tlaskalova-Hogenova H, Korinek V. Troy, a tumor necrosis factor receptor family member, interacts with lgr5 to inhibit wnt signaling in intestinal stem cells. Gastroenterology. 2013;144:381–91.
39. Stange DE, Koo BK, Huch M, Sibbel G, Basak O, Lyubimova A, Kujala P, Bartfeld S, Koster J, Geahlen JH, Peters PJ, van Es JH, van de Wetering M, Mills JC, Clevers H. Differentiated Troy+ chief cells act as reserve stem cells to generate all lineages of the stomach epithelium. Cell. 2013;155:357–68.
40. Rodriguez GC, Berchuck A, Whitaker RS, Schlossman D, Clarke-Pearson DL, Bast RC Jr. Epidermal growth factor receptor expression in normal ovarian epithelium and ovarian cancer. II. Relationship between receptor expression and response to epidermal growth factor. Am J Obstet Gynecol. 1991;164:745–50.
41. Maruo T, Ladines-Llave CA, Samoto T, Matsuo H, Manalo AS, Ito H, Mochizuki M. Expression of epidermal growth factor and its receptor in the human ovary during follicular growth and regression. Endocrinology. 1993;132:924–31.
42. Ahmed N, Maines-Bandiera S, Quinn MA, Unger WG, Dedhar S, Auersperg N. Molecular pathways regulating EGF-induced epithelio-mesenchymal transition in human ovarian surface epithelium. Am J Physiol Cell Physiol. 2006;290:C1532–42.
43. Doraiswamy V, Parrott JA, Skinner MK. Expression and action of transforming growth factor alpha in normal ovarian surface epithelium and ovarian cancer. Biol Reprod. 2000;63:789–96.
44. Jindal SK, Snoey DM, Lobb DK, Dorrington JH. Transforming growth factor alpha localization and role in surface epithelium of normal human ovaries and in ovarian carcinoma cells. Gynecol Oncol. 1994;53:17–23.
45. Parrott JA, Mosher R, Kim G, Skinner MK. Autocrine interactions of keratinocyte growth factor, hepatocyte growth factor, and kit-ligand in the regulation of normal ovarian surface epithelial cells. Endocrinology. 2000;141:2532–9.
46. Parrott JA, Skinner MK. Expression and action of hepatocyte growth factor in human and bovine normal ovarian surface epithelium and ovarian cancer. Biol Reprod. 2000;62:491–500.
47. Parrott JA, Kim G, Mosher R, Skinner MK. Expression and action of keratinocyte growth factor (KGF) in normal ovarian surface epithelium and ovarian cancer. Mol Cell Endocrinol. 2000;167:77–87.
48. Nilsson E, Doraiswamy V, Parrott JA, Skinner MK. Expression and action of transforming growth factor beta (TGFbeta1, TGFbeta2, TGFbeta3) in normal bovine ovarian surface epithelium and implications for human ovarian cancer. Mol Cell Endocrinol. 2001;182:145–55.
49. Henriksen R, Gobl A, Wilander E, Oberg K, Miyazono K, Funa K. Expression and prognostic significance of TGF-beta isotypes, latent TGF-beta 1 binding protein, TGF-beta type I and type II receptors, and endoglin in normal ovary and ovarian neoplasms. Lab Investig. 1995;73:213–20.
50. Bayne RA, Donnachie DJ, Kinnell HL, Childs AJ, Anderson RA. BMP signalling in human fetal ovary somatic cells is modulated in a gene-specific fashion by GREM1 and GREM2. Mol Hum Reprod. 2016;22(9):622–33. https://doi.org/10.1093/molehr/gaw044. Epub 2016 Jul 6
51. Lamouille S, Xu J, Derynck R. Molecular mechanisms of epithelial-mesenchymal transition. Nat Rev Mol Cell Biol. 2014;15:178–96.
52. Shepherd TG, Nachtigal MW. Identification of a putative autocrine bone morphogenetic protein-signaling pathway in human ovarian surface epithelium and ovarian cancer cells. Endocrinology. 2003;144:3306–14.

53. Erickson GF, Shimasaki S. The spatiotemporal expression pattern of the bone morphogenetic protein family in rat ovary cell types during the estrous cycle. Reprod Biol Endocrinol. 2003;1:9.
54. Piek JM, van Diest PJ, Zweemer RP, Jansen JW, Poort-Keesom RJ, Menko FH, Gille JJ, Jongsma AP, Pals G, Kenemans P, Verheijen RH. Dysplastic changes in prophylactically removed Fallopian tubes of women predisposed to developing ovarian cancer. J Pathol. 2001;195:451–6.
55. Crum CP, Drapkin R, Kindelberger D, Medeiros F, Miron A, Lee Y. Lessons from BRCA: the tubal fimbria emerges as an origin for pelvic serous cancer. Clin Med Res. 2007;5:35–44.
56. Medeiros F, Muto MG, Lee Y, Elvin JA, Callahan MJ, Feltmate C, Garber JE, Cramer DW, Crum CP. The tubal fimbria is a preferred site for early adenocarcinoma in women with familial ovarian cancer syndrome. Am J Surg Pathol. 2006;30:230–6.
57. Orsulic S, Li Y, Soslow RA, Vitale-Cross LA, Gutkind JS, Varmus HE. Induction of ovarian cancer by defined multiple genetic changes in a mouse model system. Cancer Cell. 2002;1:53–62.
58. Connolly DC, Bao R, Nikitin AY, Stephens KC, Poole TW, Hua X, Harris SS, Vanderhyden BC, Hamilton TC. Female mice chimeric for expression of the simian virus 40 TAg under control of the MISIIR promoter develop epithelial ovarian cancer. Cancer Res. 2003;63:1389–97.
59. Flesken-Nikitin A, Choi KC, Eng JP, Shmidt EN, Nikitin AY. Induction of carcinogenesis by concurrent inactivation of p53 and Rb1 in the mouse ovarian surface epithelium. Cancer Res. 2003;63:3459–63.
60. Mullany LK, Fan HY, Liu Z, White LD, Marshall A, Gunaratne P, Anderson ML, Creighton CJ, Xin L, Deavers M, Wong KK, Richards JS. Molecular and functional characteristics of ovarian surface epithelial cells transformed by KrasG12D and loss of Pten in a mouse model in vivo. Oncogene. 2011;30:3522–36.
61. Clark-Knowles KV, Garson K, Jonkers J, Vanderhyden BC. Conditional inactivation of Brca1 in the mouse ovarian surface epithelium results in an increase in preneoplastic changes. Exp Cell Res. 2007;313:133–45.
62. Wu R, Hendrix-Lucas N, Kuick R, Zhai Y, Schwartz DR, Akyol A, Hanash S, Misek DE, Katabuchi H, Williams BO, Fearon ER, Cho KR. Mouse model of human ovarian endometrioid adenocarcinoma based on somatic defects in the Wnt/beta-catenin and PI3K/Pten signaling pathways. Cancer Cell. 2007;11:321–33.
63. Wu R, Zhai Y, Kuick R, Karnezis AN, Garcia P, Naseem A, TC H, Fearon ER, Cho KR. Impact of oviductal versus ovarian epithelial cell of origin on ovarian endometrioid carcinoma phenotype in the mouse. J Pathol. 2016;240:341–51.
64. Szabova L, Yin C, Bupp S, Guerin TM, Schlomer JJ, Householder DB, Baran ML, Yi M, Song Y, Sun W, McDunn JE, Martin PL, Van Dyke T, Difilippantonio S. Perturbation of Rb, p53, and Brca1 or Brca2 cooperate in inducing metastatic serous epithelial ovarian cancer. Cancer Res. 2012;72:4141–53.
65. Sherman-Baust CA, Kuhn E, Valle BL, Shih Ie M, Kurman RJ, Wang TL, Amano T, Ko MS, Miyoshi I, Araki Y, Lehrmann E, Zhang Y, Becker KG, Morin PJ. A genetically engineered ovarian cancer mouse model based on fallopian tube transformation mimics human high-grade serous carcinoma development. J Pathol. 2014;233:228–37.
66. Perets R, Wyant GA, Muto KW, Bijron JG, Poole BB, Chin KT, Chen JY, Ohman AW, Stepule CD, Kwak S, Karst AM, Hirsch MS, Setlur SR, Crum CP, Dinulescu DM, Drapkin R. Transformation of the fallopian tube secretory epithelium leads to high-grade serous ovarian cancer in Brca;Tp53;Pten models. Cancer Cell. 2013;24:751–65.
67. Kim J, Coffey DM, Creighton CJ, Yu Z, Hawkins SM, Matzuk MM. High-grade serous ovarian cancer arises from fallopian tube in a mouse model. Proc Natl Acad Sci U S A. 2012;109:3921–6.
68. van der Horst PH, van der Zee M, Heijmans-Antonissen C, Jia Y, DeMayo FJ, Lydon JP, van Deurzen CH, Ewing PC, Burger CW, Blok LJ. A mouse model for endometrioid ovarian cancer arising from the distal oviduct. Int J Cancer. 2014;135:1028–37.

69. Eddie SL, Quartuccio SM, Oh E, Moyle-Heyrman G, Lantvit DD, Wei JJ, Vanderhyden BC, Burdette JE. Tumorigenesis and peritoneal colonization from fallopian tube epithelium. Oncotarget. 2015;6:20500–12.
70. Nagendra PB, Goad J, Nielsen S, Rassam L, Lombard JM, Nahar P, Tanwar PS. Ovarian hormones through Wnt signalling regulate the growth of human and mouse ovarian cancer initiating lesions. Oncotarget. 2016;7:64836–53.
71. Auersperg N. The stem-cell profile of ovarian surface epithelium is reproduced in the oviductal fimbriae, with increased stem-cell marker density in distal parts of the fimbriae. Int J Gynecol Pathol. 2013;32:444–53.
72. Auersperg N. The origin of ovarian carcinomas: a unifying hypothesis. Int J Gynecol Pathol. 2011;30:12–21.
73. McNairn AJ, Guasch G. Epithelial transition zones: merging microenvironments, niches, and cellular transformation. Eur J Dermatol. 2011;21:21–8.
74. Fluhmann CF, Lyons HM. Carcinoma in situ of the uterine cervix—diagnosis by biopsy. Calif Med. 1960;92:194–7.
75. Martens JE, Smedts FM, Ploeger D, Helmerhorst TJ, Ramaekers FC, Arends JW, Hopman AH. Distribution pattern and marker profile show two subpopulations of reserve cells in the endocervical canal. Int J Gynecol Pathol. 2009;28:381–8.
76. Herfs M, Yamamoto Y, Laury A, Wang X, Nucci MR, McLaughlin-Drubin ME, Munger K, Feldman S, McKeon FD, Xian W, Crum CP. A discrete population of squamocolumnar junction cells implicated in the pathogenesis of cervical cancer. Proc Natl Acad Sci U S A. 2012;109:10516–21.
77. Arnold K, Sarkar A, Yram MA, Polo JM, Bronson R, Sengupta S, Seandel M, Geijsen N, Hochedlinger K. Sox2(+) adult stem and progenitor cells are important for tissue regeneration and survival of mice. Cell Stem Cell. 2011;9:317–29.
78. Sato T, Vries RG, Snippert HJ, van de Wetering M, Barker N, Stange DE, van Es JH, Abo A, Kujala P, Peters PJ, Clevers H. Single Lgr5 stem cells build crypt-villus structures in vitro without a mesenchymal niche. Nature. 2009;459:262–5.

Tumor-Initiating Cells in Ovarian Cancer

5

Sharmila A. Bapat

Abstract

Despite advances in clinical oncology, ovarian cancer remains a leading cause of cancer-related deaths in women. Research over the last decade suggests that the aggressive nature of this disease may be attributed to cancer stem cells (CSCs), which represent a very small fraction within tumors with highly regenerative capabilities that achieves tumor growth, invasion, and metastases. These features not only indicate similarities with normal stem cells; moreover CSCs also generate cellular heterogeneity that is now recognized to be a defining characteristic of tumors. Despite this emerging knowledge, identification of ovarian CSCs has been a major challenge due to poor understanding of the niches that lead to their enrichment and lack of complete correlation between surface marker expression and functionality. Herein, I review the current literature in the field with reference to different strategies employed for CSC isolation within the maze of tumor heterogeneity. This suggests an imperative redefinition of CSCs as tumor-initiating cells, which may be more pertinent on the background of their capabilities of regeneration, reversible quiescence, and overcoming microenvironmental stress.

Keywords

Ovarian cancer · CSCs · TICs · Surface markers · Intra-tumor heterogeneity

---

S. A. Bapat
National Centre for Cell Science, Pune, India
e-mail: sabapat@nccs.res.in

H. Katabuchi et al. (eds.), *Cell Biology of the Ovary*,
https://doi.org/10.1007/978-981-10-7941-2_5

5.1 Introduction

Despite considerable advances in cancer treatments over the last few decades, only a fraction of patients actually enter a state of long-term disease-free remission. Several variables influence these outcomes, including molecular and cellular contributions such as expression of specific drug targets, drug resistance mechanisms of tumor and peripheral cells, tumor heterogeneity, etc. that largely define the nature of residual cells in tumors following completion of therapy [1]. Such submicroscopic persistence of tumor cells in the individual following treatment often referred to as "minimal residual disease" (MRD) involves a fine balance between cellular quiescence and regeneration. Tumor dormancy and disease recurrence provide an impetus to study regenerative capabilities in heterogeneous tumors. Most therapies fail to consider differences in drug sensitivities between dormant cells and those that are actively proliferating; the former persists post-therapy as MRD and can give rise to the recurrence of drug-refractory or resistant disease in the same patient.

The quiescent yet regenerative nature associated with these cell populations led to speculation of their being stemlike cells in tumors [2]. This was supported by the notion that normal stem cells, due to their lifelong existence in an individual, are likely to silently accumulate mutations until a threshold is reached when they transform to give rise to a cancer stem cell (CSC). Such hypotheses were affirmed for the first time in leukemia through identification of tumor cells with surface markers of hematopoietic stem cells and capability of self-renewal and regeneration through establishment of a hierarchy quite similar to that generating blood cells albeit the fact that some cells in the former were maturation arrested [3]. Extending these findings to solid tumors presented more challenges since such tumors are relatively inaccessible and regenerative mechanisms of their normal counterparts were not as well elucidated at that time as the blood cell lineages. Nevertheless, the concept of CSC-driven carcinogenesis steadily gained recognition and was demonstrated a decade later in neural and breast tumors [4–6]. There efforts again relied on drawing of parallels between surface markers of normal tissue stem cells to establish clear correlation with CSCs.

5.2 Stem Cells in the Normal Ovary and in Ovarian Cancer

It increasingly also became evident that the presence of CSCs in tumors maximizes its growth and metastatic potential besides providing for MRD. However, CSC identification presented several challenges, an important one being that even today, a lack of definitive information exists regarding stem cell populations and their lineage commitment in several normal tissues. Historically, regenerative potential in the ovary was associated with the ovarian surface epithelium (OSE, often referred to as germinal epithelium; [7, 8]). Bipotent stem cells were identified in the tunica albuginea that could generate OSE, granulosa, and germ cells [9]. Dedifferentiation and proliferation of granulosa cells in culture along with progressive acquisition of mesenchymal markers not only indicated a transition from a epithelial to

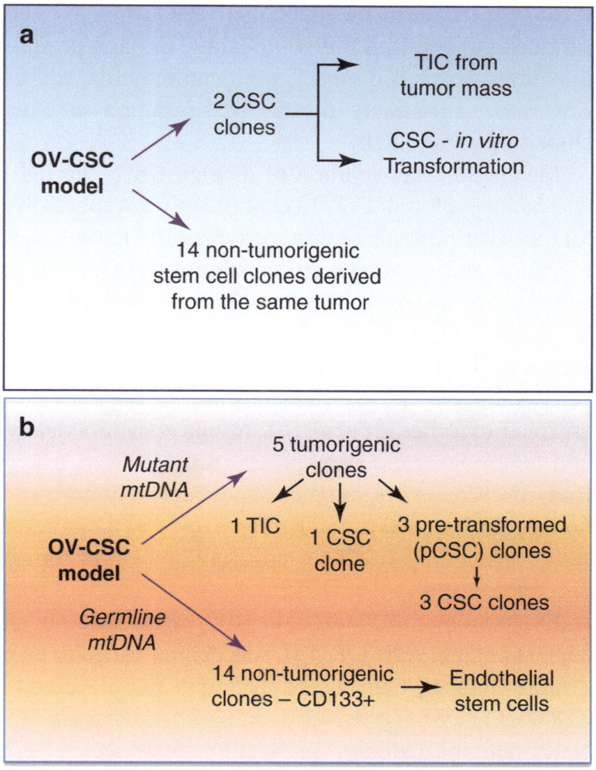

Fig. 5.1 The ovarian cancer stem cell (OV-CSC) model. (**a**) One clone was a tumor-initiating clone (TIC), another was a CSC clone generated following transformation in vitro, while the remaining 19 were non-tumorigenic. (**b**) Mitochondrial mutation (mtDNA) profiling revealed two groups in this model: one expressed a mutant mtDNA profile, comprised of 1 TIC, 1 CSC, and 3 other pre-transformed clones that like the previous CSC underwent in vitro transformation and generated CSCs, and the other group of 14 clones expressed the germline mtDNA profile and were identified to be endothelial stem cells

mesenchymal phenotype but also suggested acquisition of stem cell-like state [10]. The latter was experimentally supported by Oct4 expression and redifferentiation along the neuronal, chondrocytic, and osteoblastic lineages. More recently, co-expression of LGR5, LEF1, CD133, and CK6B with ALDH1 at the hilum (junction of the ovary, mesothelium, and fallopian tube) was further reported to identify an unrecognized stem cell niche [11]. However, it has been difficult to draw parallels between these two cell types and the existence of stem cells in ovarian tumors.

Over 15 years ago, an aggressive and highly regenerative nature of cells within primary high-grade serous ovarian tumors when brought into culture suggested the presence of stemlike cells and led to several attempts in isolation of single cell clones and testing their long-term survival in culture. One particular highly metastatic tumor ascites yielded 19 spontaneously immortalized clones, characterization of which for expression of stem cell markers and regenerative functions identified 1 tumor-initiating clone, another clone that acquired tumorigenic properties after a few passages in culture, and 17 untransformed/pre-tumorigenic stem cell clones (Fig. 5.1a; [12]). The former two clones were suggested to be CSCs by virtue of abilities of anchorage-independent growth (formation of multicellular colonies in soft agar and spheroids in suspension culture) and generated mouse tumor xenografts histopathologically similar to the human tumor. Our approach of CSC identification relied on functional readouts/assays rather than application of cell surface markers. In a follow-up study, through profiling of mitochondrial genome mutations in this cell model, we identified three additional clones (besides the two CSCs) that were likely to be derived from the same transformed parental stem cell clone (Fig. 5.1b; [13]).

Meanwhile the isolation of ovarian CSCs through the use of surface markers was being explored. CD133 expression was reportedly enriched in primary ovarian tumors over normal, benign, or metastatic tissues and correlated with higher proliferation and clonogenicity capabilities [14]. Another study not only supported CD133 to be an ovarian CSC marker but also identified its epigenetic dysregulation in association with transformation [15]. In these two studies, an association between CD133 expression and clinicopathologic features of the disease could not be established, and CD133-cells were also found to be weakly tumorigenic as affirmed by other groups [16]. In our above model system, we found that the 14 non-tumorigenic clones expressed CD133; moreover in our hands, CD133+ cells from ovarian tumors and metastases lacked tumorigenic capabilities. Further functional correlates led us to identify CD133 as a marker of primitive endothelial stem cells recruited by CSCs toward development of tumor vasculature during disease progression (Fig. 5.1b; [17]).

Expression of CD44+/CD117+ reported earlier in OSE cells, although not in the context of stem cells [18, 19], was demonstrated to be useful in isolating cells from ovarian tumors capable of forming multilayered spheroids in culture [20]. CD44+/MYD88+ cells, also proposed as ovarian CSCs, were shown to harbor NFκβ activity, cytokine production, resistance to conventional chemotherapies, tumor necrosis factor-α-mediated apoptosis, formation of suspended spheroids, etc. [21], which effectively promotes a pro-inflammatory microenvironment. The significance of these markers in aberrant CSC-driven repair/differentiation is a similarity with OSE wound healing/tissue repair processes involving TLR4-MYD88 signaling [7]. The same research group also identified bipotent CD44+/CDH1+/CD34+ cells in ovarian cancer capable of tumor regeneration as well as vascularization; these were shown to secrete proangiogenic factors IL6, IL8, and VEGF and inhibit IKK during tumor development [22]. Enrichment of stem cells in "side population" (SP) fractions that relies on active expulsion of molecules, such as the dye Hoechst 33342 in flow-associated cell sorting of mouse ovaries, ovarian cancer cell lines, and tumor ascites of patients, was also reported [8, 23–26]. SP populations were capable of self-renewal, generated descendent SP and non-SP populations, and were highly resistant to doxorubicin over the non-SP cells through ABCG2 expression, which together supports their being CSCs.

The presence of ALDH1 autoantibodies in ovarian cancer patients suggested an association with tumor cells. ALDH1 expression has since been demonstrated in cells that exhibit non-adherent growth in vitro, multidrug resistance, and in vivo tumorigenicity and is enriched in highly aggressive, poor-prognosis tumors [27–32]. The chemoresistant and metastatic behavior of Aldh1+ ovarian CSCs has been attributed with expression of specific miRNAs in these cells [33]. Co-expression of either ROR1 or musashi-1 expression with ALDH1 expression in ovarian tumors or xenografts was also detected to correlate well with CSC characteristics, higher rates of relapse, and a shorter median survival than patients with ovarian cancers that suggested them to be ancillary CSC markers [34].

Ovarian cancer cells have been suggested to harbor phenotypic plasticity, which in the normal context was earlier attributed to wound healing functions of the OSE and recently of fallopian tube epithelia [35]. This in part appears to have precluded the identification of CSC markers as in the case of leukemia and a few solid cancers wherein a definitive correlation between an immunophenotype and functional properties could be established [36]. While quiescence is conjectured to be a defining property of CSCs especially in the context of tumor dormancy, in the rush of CSC reports in various malignancies, it failed to be demonstrated. Label-retaining cell assays with vital fluorophores that are equally partitioned among daughter cells with reduced fluorescence intensity have been useful in identifying slow-cycling stem cells and enable live cell monitoring when subsequent functional analysis is required [37]. In a label-chase approach using membrane-labeling dyes (PKH67/PKH26), we demonstrated stem cell activity to be enriched in label-retaining quiescent (G0) or slow-cycling PKH^{hi} fractions of ovarian xenografts, and partial quenching identified PKH^{lo} cells as a cycling progenitor population, while a large fraction of the tumor is comprised of differentiated PKH^{neg} cells totally devoid of label [38]. Notably, PKH^{hi} cells are capable of reversing their quiescent state under specific microenvironmental triggers to exhibit a highly clonogenic and tumorigenic nature, which distinguished them from any other nonproliferative cell state in tumors. This defined a universal scheme for identification of CSCs and has been applied in several studies [39–42]. During tumor formation and maintenance, CSCs like normal stem cells continue to self-renew and generate differentiated derivatives that constitute the bulk of a tumor.

5.3 Redefining Reversible Quiescence as a Functional Property of CSCs Reappropriated Their Recognition as Tumor-Initiating Cells (TICs)

Identification of CSCs in tumors has now been extensively studied. However, in some tumor types such as melanoma, especially under conditions of stress, our understanding of tissue homeostasis cannot account for the presence of CSCs, i.e., in terms of normal stem cells serving as targets of oncogenic transformation [43]. To circumvent this limitation, it was suggested that transit-amplifying progenitors or even differentiated cells might undergo the initial oncogenic genomic rearrangements [44]. This implies dedifferentiation, which would necessitate concurrent acquisition of self-renewal capabilities and maturation defects associated with cancer [45], and may not always involve acquisition of the same immunophenotype(s) as a tissue stem cell. From the perspective of aberrant protein expression being a biochemical hallmark of cancer, it also seems unrealistic to identify identical lineage-defining normal stem cell molecular markers following transformation when tissue homeostasis is known to be perturbed. It thereby becomes improbable to assign a specific immunophenotype to cells with differing capabilities of self-renewal, differentiation, and regeneration in a tumor that increasingly becomes heterogeneous and acquires an aggressive nature [46].

Stemlike cells in tumors exhibiting the properties of self-renewal, drug resistance, and tumor regeneration are thus believed to arise from either:

1. Feed-forward mechanisms—aberrant tissue regeneration from transformed stem cells
2. Feedback mechanisms—dedifferentiation as an adaptive mechanism to overcome microenvironmental stress

The above resolution thereby emphasizes a subtle yet demarcation between a tumor-initiating cell (TIC) and a CSC: the former being assigned proliferative and regenerative functions and the latter being capable of self-renewal and multipotency. Both cells are likely to exist in tumors, and their phenotypes and cross talks may change dynamically in response to environmental cues [47]. The frequency of TICs could thus vary from being relatively rare to comprising a significant fraction of tumors, implying that establishment of mitotic quiescence than expression of "stemness" features is a more important characteristic of CSCs [36].

5.4 CSC/TIC Identification Through the Maze of Intra-Tumor Heterogeneity and Their Differential Regulation Under Conditions of Stress

Cancer progression is purportedly aligned with Darwinian evolution that involves continual genetic instability to enhance cellular complexity and increase avenues of survival by generating diverse clones within the same tumor that may reacquire stemlike properties [48]. While the extensive use of colony-forming assays, limiting dilution transplantation, and lineage-tracing assays has helped establish regenerative potential in tumors [49], it is realized that they do not account for the entire range of intra-tumor heterogeneity. We had earlier identified and isolated aneuploid cells from experimental tumors that remain dormant in conditions of steady-state growth but, when subjected to environmental stress such as limiting nutrients or therapy, provide an adaptive relief to the threatened tumor existence by reentering into the cell cycle and proliferating in a drug-resistant manner. Reacquisition of stemlike properties by these cells further leads to the development of hierarchies parallel to that of canonical CSCs. This further leads to the coexistence of discrete CSC pools and reflects on emerging heterogeneity within the same tumor. To address these basic questions in the field today, it becomes necessary to map out all possible correlates of the regenerative hierarchy, genetic instability, and stimuli for clonal selection and differential cycling along with the diversity of molecular profiles within an organ/tumor.

Multiparameter flow cytometry has been useful in elucidation of cellular and molecular heterogeneity within tumors [50]. Using these techniques, we extended our label-chase approach of label-chase-based resolution of the regenerative hierarchy to include a second level of detection of distinct host, euploid and aneuploid fractions based on their distinct DNA content(s), and yet another level that identified the different phases of the cell cycle (Fig. 5.2). Integration of all 3 levels identifies 18 discrete cell populations [51]. Each one of these cell fractions has been evaluated for regenerative potential under steady-state conditions as well as on exposure to several drugs including mitotic inhibitors, epigenetic modulators, etc. Remarkably, we found that each drug effectively eliminated cells expressing its specific molecular target(s); however, drug-resistant/drug-refractory residual fractions that evaded single drug treatment led to generation of parallel hierarchies and repopulation of the tumor. A common feature observed with all drugs was drastic reduction in all the PKH^{neg} (differentiated) subpopulations that form the bulk of the tumor and thereby significantly reduced tumor volumes that suggest effective tumor targeting. However, residual CSCs/PKH^{lo} progenitors initiate tumor repopulation; additionally, it was observed that genetic diversity of aneuploid tumor populations further provides for selection of specific resistant clone(s) in response to each drug. Different populations are thus likely to contribute to disease recurrence depending on the composition of the therapeutic regime.

Under such circumstances, complete remission would necessitate formulation of drug combinations depending on the targeted populations and an understanding of mechanistic action of these drugs. While we evaluated several combinations, the most effective appeared to be one that included gemcitabine (currently used in ovarian cancer treatment and known to target RRM2) along with 5Aza-dC (a demethylating agent for which reexpression of a tumor suppressor PTGIS is an effective biomarker; [52]) and a cytotoxic monoclonal antibody developed in the lab (termed mAb150 that recognizes the cells expressing the antigen ANXA2). Gemcitabine was demonstrated to target most of the proliferating tumor cells, 5Aza-dC importantly also restricted growth of aneuploid populations, while mAb150 was specifically effective in mediating apoptosis in CSC and progenitor fractions. The contribution of 5Aza-dC was particularly remarkable since it also

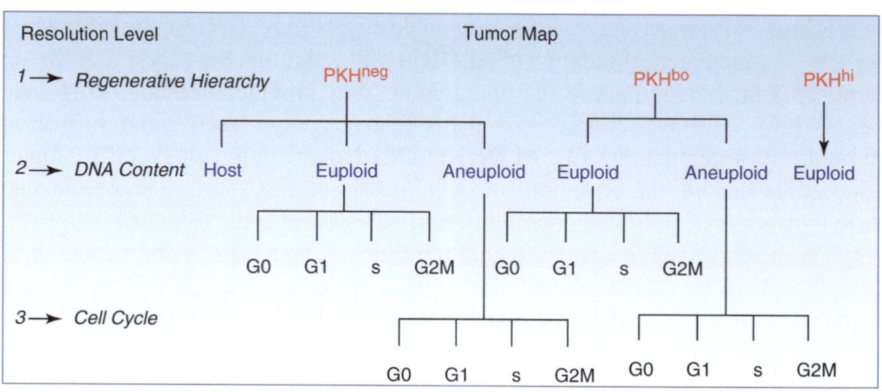

Fig. 5.2 Resolution of intra-tumor heterogeneity based on level 1, regenerative hierarchy; level 2, DNA content; level 3, cell cycle phases

enhanced RRM2 and ANXA2 expression in cells wherein the CpG islands in their promoters were methylated, thereby leading to an increased frequency of RRM2+ and ANXA2+ cells in the tumor that could further be recognized and targeted by the other two drugs. This complementation of targets and mechanism of drug action led to total tumor regression following a six-dose regime and is an example in point of not only eliminating all possible active and latent regenerative cell populations in tumors.

5.5 Future Perspectives

The realization of epithelial ovarian cancer as a stem cell disease has the potential to dramatically change approaches in its treatment. While until a few years ago the understanding of these cells was poorly defined and their identification remained elusive, recent studies have explored various approaches and themes to prospectively isolate these cells from cell models and tumor samples. The purported existence of TICs along with CSCs in tumors has further improved our comprehension of the emergence of various subtypes from a single type of transformed cell. There are still important unanswered questions, addressing of which will lead to new insights into the functions of TICs in cancer toward the design of novel therapeutic strategies. While most experiments have been conducted in animal models, recent attempts emphasize on the imminent necessity of development of methods for effective evaluation of potential MRD in patients [53].

Continuing such studies toward characterization of progenitors, CSCs and TICs, and genetic instability in drug-resistant ovarian cancer is likely to lead to a greater understanding of early events leading to the genesis of MRD, in addition to providing new therapeutic targets aimed at the elimination of cells directly responsible for MRD. An ovarian cancer-specific CD117-CD44-based vaccine in athymic nude mice that induced enhanced serum IFN-γ, decreased TGF-β levels, increased cytotoxic activity of natural killer cells, and reduced CD117+/CD44+ as well as ALDH1+ CSCs is reported [54]. mTOR signaling, Her-2/neu, CD47, and FGF18/FGFR3 have also been suggested as druggable targets in ALDH1+ cells [55], while the VEGFA-mediated stimulation of ALDH1+ CSCs through Src-DNMT3A-driven *miR-128-2* methylation and Bmi1 upregulation may provide alternative druggable targets [56]. All-trans-retinoic acid also effectively suppresses tumor formation through downregulation of ALDH1/FoxM1/Notch1 signaling [57]. Other approaches include the development of inhibitors of CSC surface molecule like drug transporters, specific oncoproteins, normal and neoplastic stem cell pathways, survival factors, oxidative stress, signal transduction pathways, niche factors, epigenetic and posttranslational regulation of gene expression, etc. [58]. Conclusively, while recent findings have the potential to positively influence the outcome of ovarian cancer patients, multifaceted therapeutic approaches targeting CSCs and/or TICs are likely to be a mainstay of the future.

References

1. Shimizu T, Sugihara E, Yamaguchi-Iwai S, Tamaki S, Koyama Y, Kamel W, Ueki A, Ishikawa T, Chiyoda T, Osuka S, Onishi N, Ikeda H, Kamei J, Matsuo K, Fukuchi Y, Nagai T, Toguchida J, Toyama Y, Muto A, Saya H. IGF2 preserves osteosarcoma cell survival by creating an autophagic state of dormancy that protects cells against chemotherapeutic stress. Cancer Res. 2014;74:6531–41.
2. Zakaria N, Satar NA, Abu Halim NH, Ngalim SH, Yusoff NM, Lin J, Yahaya BH. Targeting lung cancer stem cells: research and clinical impacts. Front Oncol. 2017;5:80.
3. Bonnet D, Dick JE. Human acute myeloid leukemia is organized as a hierarchy that originates from a primitive hematopoietic cell. Nat Med. 1997;3:730–7.
4. Al-Hajj M, Wicha MS, Benito-Hernandez A, Morrison SJ, Clarke MF. Prospective identification of tumorigenic breast cancer cells. Proc Natl Acad Sci U S A. 2003;100:3983–8.
5. Singh SK, Clarke ID, Terasaki M, Bonn VE, Hawkins C, Squire J, Dirks PB. Identification of a cancer stem cell in human brain tumors. Cancer Res. 2003;63:5821–8.
6. Xia P. Surface markers of cancer stem cells in solid tumors. Curr Stem Cell Res Ther. 2014;9:102–11.
7. Auersperg N, Wong AS, Choi KC, Kang SK, Leung PC. Ovarian surface epithelium: biology, endocrinology, and pathology. Endocr Rev. 2001;22:255–88.
8. Szotek PP, Chang HL, Brennand K, Fujino A, Vanmarcke RP, Celso CL, Dombkowski D, Preffer F, Cohen KS, Teixeira J, Donahoe PK. Normal ovarian surface epithelial label-retaining cells exhibit stem/progenitor cell characteristics. Proc Natl Acad Sci U S A. 2008;105:12469–73.
9. Bukovsky A, Caudle MR, Svetlikova M, Upadhyaya NB. Origin of germ cells and formation of new primary follicles in adult human ovaries. Reprod Biol Endocrinol. 2004;2:20–10.
10. Kossowska-Tomaszczuk K, Geyter De C, Geyter De M, Martin I, Holzgreve W, Scherberich A, Zhang H. The multipotency of luteinizing granulosa cells collected from mature ovarian follicles. Stem Cells. 2009;27:210–9.
11. Flesken-Nikitin A, Hwang CI, Cheng CY, Michurina TV, Enikolopov G, Nikitin AY. Ovarian surface epithelium at the junction area contains cancer-prone stem cell niche. Nature. 2013;495:241–5.
12. Bapat SA, Mali AM, Koppikar CB, Kurrey NK. Stem and progenitor-like cells contribute to the aggressive behavior of human epithelial ovarian cancer. Cancer Res. 2005;65:3025–9.
13. Wani AA, Sharma N, Shouche YS, Bapat SA. Nuclear-mitochondrial genomic profiling reveals a pattern of evolution in epithelial ovarian tumor stem cells. Oncogene. 2006;25:6336–44.
14. Ferrandina G, Bonanno G, Pierelli L, Perillo A, Procoli A, Mariotti A, Corallo M, Martinelli E, Rutella S, Paglia A, Zannoni G, Mancuso S, Scambia G. Expression of CD133-1 and CD133-2 in ovarian cancer. Int J Gynecol Cancer. 2008;18:506–14.
15. Baba T, Convery PA, Matsumura N, Whitaker RS, Kondoh E, Perry T, Huang Z, Bentley RC, Mori S, Fujii S, Marks JR, Berchuck A, Murphy SK. Epigenetic regulation of CD133 and tumorigenicity of CD133+ ovarian cancer cells. Oncogene. 2009;28:209–18.
16. Curley MD, Therrien VA, Cummings CL, Sergent PA, Koulouris CR, Friel AM, Roberts DJ, Seiden MV, Scadden DT, Rueda BR, Foster R. CD133 expression defines a tumor initiating cell population in primary human ovarian cancer. Stem Cells. 2009;27:2875–83.
17. Kusumbe AP, Mali AM, Bapat SA. CD133-expressing stem cells associated with ovarian metastases establish an endothelial hierarchy and contribute to tumor vasculature. Stem Cells. 2009;27:498–508.
18. Cannistra SA, Abu-Jawdeh G, Niloff J, Strobel T, Swanson L, Andersen J, Ottensmeier C. CD44 variant expression is a common feature of epithelial ovarian cancer: lack of association with standard prognostic factors. J Clin Oncol. 1995;13:1912–21.
19. Parrott JA, Kim G, Skinner MK. Expression and action of kit ligand/stem cell factor in normal human and bovine ovarian surface epithelium and ovarian cancer. Biol Reprod. 2000;62:1600–9.

20. Zhang S, Balch C, Chan MW, Lai HC, Matei D, Schilder JM, Yan PS, Huang TH, Nephew KP. Identification and characterization of ovarian cancer-initiating cells from primary human tumors. Cancer Res. 2008;68:4311–20.
21. Alvero AB, Chen R, Fu HH, Montagna M, Schwartz PE, Rutherford T, Silasi DA, Steffensen KD, Waldstrom M, Visintin I, Mor G. Molecular phenotyping of human ovarian cancer stem cells unravels the mechanisms for repair and chemoresistance. Cell Cycle. 2009a;8:158–66.
22. Alvero AB, Fu HH, Holmberg J, Visintin I, Mor L, Marquina CC, Oidtman J, Silasi DA, Mor G. Stem-like ovarian cancer cells can serve as tumor vascular progenitors. Stem Cells. 2009b;27:2405–13.
23. Ambler CA, Maatta A. Epidermal stem cells: location, potential and contribution to cancer. J Pathol. 2009;217:206–16.
24. Gao Q, Geng L, Kvalheim G, Gaudernack G, Suo Z. Identification of cancer stemlike side population cells in ovarian cancer cell line OVCAR-3. Ultrastruct Pathol. 2009;33:175–81.
25. Lin KK, Goodell MA. Purification of hematopoietic stem cells using the side population. Methods Enzymol. 2006;420:255–64.
26. Szotek PP, Pieretti-Vanmarcke R, Masiakos PT, Dinulescu DM, Connolly D, Foster R, Dombkowski D, Preffer F, Maclaughlin DT, Donahoe PK. Ovarian cancer side population defines cells with stem cell-like characteristics and Mullerian inhibiting substance responsiveness. Proc Natl Acad Sci U S A. 2006;103:11154–9.
27. Ayub TH, Keyver-Paik M-D, Debald M, Rostamzadeh B, Thiesler T, Schröder L, Barchet W, Abramian A, Kaiser C, Kristiansen G, Kuhn W, Kubler K. Accumulation of ALDH1-positive cells after neoadjuvant chemotherapy predicts treatment resistance and prognosticates poor outcome in ovarian cancer. Oncotarget. 2015;6:16437–48.
28. Kakar SS, Worth CA, Wang Z, Carter K, Ratajczak M, Gunjal P. DOXIL when combined with Withaferin a (WFA) targets ALDH1 positive cancer stem cells in ovarian cancer. J Cancer Stem Cell Res. 2016;4:e1002.
29. Kuroda T, Hirohashi Y, Torigoe T, Yasuda K, Takahashi A, Asanuma H, et al. ALDH1-high ovarian cancer stem-like cells can be isolated from serous and clear cell adenocarcinoma cells, and ALDH1 high expression is associated with poor prognosis. PLoS One. 2013;8:e65158.
30. Liebscher CA, Prinzler JS, Sinn BV, Budczies J, Denkert C, Noske A, Sehouli J, Loana E. Aldehyde dehydrogenase 1/epidermal growth factor receptor coexpression is characteristic of a highly aggressive, poor-prognosis subgroup of high-grade serous ovarian carcinoma. Hum Pathol. 2013;44:1465–71.
31. Penumatsa K, Edassery SL, Barua A, Bradaric MJ, Luborsky JL. Differential expression of aldehyde dehydrogenase 1a1 (ALDH1) in normal ovary and serous ovarian tumors. J Ovarian Res. 2010;3:28.
32. Pylvas-Eerola M, Liakka A, Puistola U, Koivunen J, Karihtala P. Cancer stem cell properties as factors predictive of chemoresistance in neoadjuvantly-treated patients with ovarian cancer. Anticancer Res. 2016;36:3425–31.
33. Park YT, Jeong JY, Lee MJ, Kim KI, Kim TH, et al. MicroRNAs overexpressed in ovarian ALDH1-positive cells are associated with chemoresistance. J Ovarian Res. 2013;6:18.
34. Chen J, Xia Q, Jiang B, Chang W, Yuan W, Ma Z, et al. Prognostic value of cancer stem cell marker ALDH1 expression in colorectal cancer: a systematic review and meta-analysis. PLoS One. 2015;10:e0145164.
35. Garson K, Vanderhyden BC. Epithelial ovarian cancer stem cells: underlying complexity of a simple paradigm. Reproduction. 2015;149:59–70.
36. Wang L, Cao J, Wang Y, Lan T, Liu L, Wang W, Jin N, Gong J, Zhang C, Teng F, Yan G, Li C, Li J, Wan H, Hu B, Li W, Zhao X, Qi Z, Zhao T, Zhou Q. Immunogenicity and functional evaluation of iPSC-derived organs for transplantation. Cell Discov. 2015;1:15015.
37. Yue Z, Jiang TX, Widelitz RB, Chuong CM. Mapping stem cell activities in the feather follicle. Nature. 2005;438:1026–9.
38. Kusumbe AP, Bapat SA. Cancer stem cells and aneuploid populations within developing tumors are the major determinants of tumor dormancy. Cancer Res. 2009;69:9245–53.

39. Akrap N, Andersson D, Bom E, Gregersson P, Stahlberg A, Landberg G. Identification of distinct breast cancer stem cell populations based on single-cell analyses of functionally enriched stem and progenitor pools. Stem Cell Rep. 2016;6:121–36.
40. Moro M, Bertolini G, Pastorino U, Roz L, Sozzi G. Combination treatment with all-trans retinoic acid prevents Cisplatin-induced enrichment of CD133+ tumor-initiating cells and reveals heterogeneity of cancer stem cell compartment in lung cancer. J Thorac Oncol. 2015;10:1027–36.
41. Richichi C, Brescia P, Alberizzi V, Fornasari L, Pelicci G. Marker-independent method for isolating slow-dividing cancer stem cells in human glioblastoma. Neoplasia. 2013;15:840–7.
42. Zeuner A, Francescangeli F, Contavalli P, Zapparelli G, Apuzzo T, Eramo A, Baiocchi M, De Angelis ML, Biffoni M, Sette G, Todaro M, Stassi G, De Maria R. Elimination of quiescent/slow-proliferating cancer stem cells by Bcl-XL inhibition in non-small cell lung cancer. Cell Death Differ. 2014;21:1877–88.
43. Quintana E, Shackleton M, Sabel MS, Fullen DR, Johnson TM, Morrison SJ. Efficient tumour formation by single human melanoma cells. Nature. 2008;8:456–593.
44. Prasetyanti PR, Medema JP. Intra-tumor heterogeneity from a cancer stem cell perspective. Mol Cancer. 2017;16:41.
45. Quintana E, Shackleton M, Foster HR, Fullen DR, Sabel MS, Johnson TM, Morrison SJ. Phenotypic heterogeneity among tumorigenic melanoma cells from patients that is reversible and not hierarchically organized. Cancer Cell. 2010;18:510–23.
46. Rehe K, Wilson K, Bomken S, et al. Acute B lymphoblastic leukaemia propagating cells are present at high frequency in diverse lymphoblast populations. EMBO Mol Med. 2013;5:38–51.
47. Boiko AD, Razorenova OV, Rijn M, Swetter SM, Johnson DL. Human melanoma-initiating cells express neural crest nerve growth factor receptor CD271. Nature. 2010;466:133–7.
48. Mroz EA, Rocco JW. The challenges of tumor genetic diversity. Cancer. 2017;123:917–27.
49. Rycaj K, Tang DG. Cell-of-origin of cancer versus cancer stem cells: assays and interpretations. Cancer Res. 2015;75:4003–11.
50. Di Palma S, Bodenmiller B. Unraveling cell populations in tumors by single-cell mass cytometry. Curr Opin Biotechnol. 2015;31:122–9.
51. Naik RR, Singh AK, Mali AM, Khirade MF, Bapat SA. A tumor deconstruction platform identifies definitive end points in the evaluation of drug responses. Oncogene. 2016;35:727–37.
52. Singh AK, Chandra N, Bapat SA. Evaluation of epigenetic drug targeting of heterogenous tumor cell fractions using potential biomarkers of response in ovarian cancer. Clin Cancer Res. 2015;21:5151–63.
53. Mosna F, Capelli D, Gottardi M. Minimal residual disease in acute myeloid leukemia: still a work in progress. J Clin Med. 2017;6:1–4.
54. Wu D, Wang J, Cai Y, Ren M, Zhang Y, Shi F, Zhao F, He X, Pan M, Yan C, Dou J. Effect of targeted ovarian cancer immunotherapy using ovarian cancer stem cell vaccine. J Ovarian Res. 2015;8:68.
55. Sharrow AC, Perkins B, Collector MI, Yu W, Simons BW, Jones RJ. Characterization of aldehyde dehydrogenase 1 high ovarian cancer cells: towards targeted stem cell therapy. Gynecol Oncol. 2016;142:341–8.
56. Jang K, Kim M, Gilbert CA, Simpkins F, Ince TA, Slingerland JM. VEGFA activates an epigenetic pathway upregulating ovarian cancer-initiating cells. EMBO Mol Med. 2017;9:304–18.
57. Young MJ, YH W, Chiu WT, Weng TY, Huang YF, Chou CY. All-trans retinoic acid downregulates ALDH1-mediated stemness and inhibits tumour formation in ovarian cancer cells. Carcinogenesis. 2015;36:498–507.
58. Tu LC, Foltz G, Lin E, Hood L, Tian Q. Targeting stem cells-clinical implications for cancer therapy. Curr Stem Cell Res Ther. 2009;4:147–53.

6 Emerging Role of CD44 Variant 6 in Driving the Metastatic Journey of Ovarian Cancer Stem Cells

Takeshi Motohara and Hidetaka Katabuchi

Abstract

In recent history, cancer stem cells have been postulated as the driving force for carcinogenesis and the seeds of metastasis in several types of solid cancers. In addition, accumulating evidence suggests that the cell adhesion molecule CD44 and its variant isoforms are critical players in regulating the biology of cancer stem cells and cancer metastasis. However, the functional role of these molecules in ovarian cancer stem cells, particularly in the development of metastasis, remains completely unresolved in ovarian cancer patients.

In this review, we highlight the emerging role of CD44, particularly CD44 variant 6, in driving peritoneal dissemination and distant parenchymal metastasis with respect to ovarian cancer stem cells. Furthermore, we discuss the potential clinical implications and a vision for the future of therapeutic strategies targeting cancer stem cells and their niche microenvironment in patients with ovarian cancer.

Keywords

Ovarian cancer · Cancer stem cell · Cancer stem cell niche · Dissemination · Metastasis · CD44 · CD44 variant 6

T. Motohara (✉) · H. Katabuchi
Department of Obstetrics and Gynecology, Faculty of Life Sciences, Kumamoto University, Kumamoto, Japan
e-mail: kan@kumamoto-u.ac.jp

H. Katabuchi et al. (eds.), *Cell Biology of the Ovary*,
https://doi.org/10.1007/978-981-10-7941-2_6

6.1 Introduction

Epithelial ovarian cancer has the highest fatality rate and is the leading cause of death among the malignancies of female genital tract [1–3]. Despite ongoing efforts to organize effective screening strategies [4, 5], a limited number of ovarian cancer patients are diagnosed before the cancer has spread outside of the ovaries (stage I). At this stage, the majority of patients can be cured with currently available treatments. However, after the disease has metastasized to the pelvic organs, including the uterus or fallopian tube (stage II), beyond the pelvis to the abdomen or nearby lymph nodes (stage III), or beyond the peritoneal cavity (stage IV), the cure rate after treatment with extirpative surgery and intensive chemotherapy decreases substantially [6–8].Although most patients with advanced ovarian cancer initially respond well to a combination of taxane- and platinum-based chemotherapy, small numbers of chemoresistant disseminated cancer cells in metastatic sites can persist and remain dormant for prolonged periods, leading to relapse [8–11].

Over a century ago, Stephen Paget proposed the "seed and soil" hypothesis, which stated that metastasis depends on cross talk between selected cancer cells (the seeds) and the proper organ microenvironment (the soil) [12, 13]. Thereafter, the study of cancer metastasis on the systemic, cellular, and molecular levels has grown exponentially, but the hallmarks of metastasis still remain completely unresolved. Thus, elucidating the mechanisms that govern metastasis may provide potential therapeutic targets for treatment of cancer patients.

During the past couple of decades, the "cancer stem cell" theory has postulated that a bulk of tumor is generated by rare populations of self-renewing, multipotent, and tumor-initiating cells, conceptually termed cancer stem cells, which are inherently responsible for tumor metastasis and therapy resistance [14–16]. In ovarian cancer, the first evidence for the existence of cancer stem cells was reported by Bapat et al. in 2005 [17], and since then, growing data have provided substantial evidence for the existence of ovarian cancer stem cells [18–20]. Although ovarian cancer stem cells have not been completely evaluated, these cells seem to play a crucial role in driving disseminated metastasis, metastatic colonization, and recurrence at peritoneal or distant metastatic sites [20, 21]. In addition, at the clinical level of ovarian cancer treatment, cancer stem cells are able to survive conventional chemotherapies, which generally target rapidly dividing cancer progenitor cells, and these small populations of cancer stem cells give rise to chemoresistant recurrent tumors in the metastatic regions [14, 21, 22] (Fig. 6.1).

Normal stem cells reside in a proper microenvironment called the "stem cell niche" that maintains their properties of stemness [23]. Accumulating evidence indicates that cancer stem cells also depend on a similar niche, dubbed the "cancer stem cell niche" [24–27], which is involved in regulating cancer stem cell features in each defined microenvironment [24, 25, 27]. For ovarian cancer stem cells, the pelvic peritoneum, covered by mesothelial cells, might provide such a specialized niche

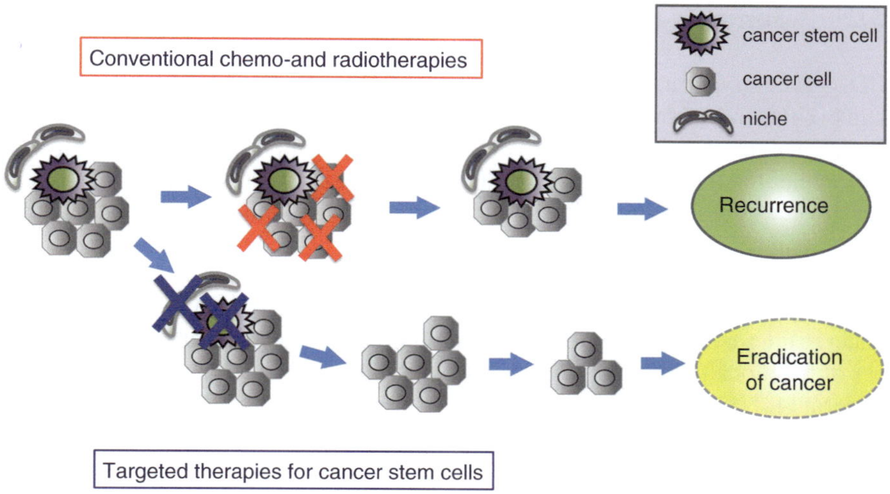

Fig. 6.1 Therapeutic strategies targeting cancer stem cells. Accumulating evidence indicates that cancer stem cells can survive conventional therapies, which mainly target cancer progenitor cells or more differentiated cancer cells with highly proliferative potential. Conversely, if novel targeted therapies against cancer stem cells are established, they may kill the self-renewing cancer stem cells more effectively and lead to the eventual extirpation of cancer

microenvironment [28]. Taken together, these fascinating facts increase the possibility that the "seeds" are one of the subpopulations of cancer stem cells and the "soil" is the specific cancer stem cell niche to maintain their cancer stemness, including the initiation of tumorigenesis and metastasis [29]. Therefore, uncovering the molecular events underlying the relationship between ovarian cancer stem cells and the cancer stem cell niche in metastatic sites is the final frontier in ovarian cancer biology.

The cell adhesion molecule CD44 is a transmembrane glycoprotein that is implicated in cell–cell and cell–matrix interactions and correlates with the malignant process, especially in cancer metastasis [30–33]. CD44 exists as a standard (CD44s) isoform and in addition as numerous variant (CD44v) isoforms generated through alternative mRNA splicing [33, 34]. It should be noted that, for more than a decade, CD44 has been rediscovered as a putative cancer stem cell marker in several types of malignant tumors, such as breast, pancreatic, gastric, colorectal, bladder, and ovarian cancers [15, 21, 35–37]. Furthermore, among CD44v isoforms, in particular CD44v6 has been reported to be associated with carcinogenesis, proliferation, and metastasis in a variety of solid cancers, including ovarian cancer [33, 38–41].

This chapter aims to demonstrate the functional role of CD44, especially CD44v6, in ovarian cancer stem cells, and we review the principles of our current understanding of these molecules in correlation with cancer stem cell properties, notably those driving peritoneal dissemination and distant metastasis in patients with ovarian cancer. Finally, we discuss the potential clinical implications and directions for future cancer stem cell research.

6.2 Genomic and Protein Structure of CD44

CD44 is a multi-structural and multifunctional complex transmembrane glycoprotein [42–44]. Members of the CD44 family participate in various physiological processes, including hematopoiesis, regeneration, and lymphocyte homing [43, 45]. More importantly, they are also implicated in a variety of pathological processes, such as cell adhesion, migration, invasion, angiogenesis, and tumor metastasis, particularly in cancer tissues [30, 46].

CD44 is encoded by the highly conserved CD44 gene on chromosome 11 [32] and consists of 20 exons in the human genome [33]. Transcripts for the CD44 gene undergo alternative splicing in various combinations to generate an enormous number of functionally different isoforms, such as the CD44s isoform and each CD44v isoform [44] (Fig. 6.2). The smallest CD44s is generated from exons 1–5 and exons 16–20, whereas a multitude of CD44v is produced by insertion of variant exons (exons 6–15), distinctively identified as v1–v10 at the membrane-proximal extracellular region [44, 47]. These CD44v may contain one or more variant regions, such as CD44v3, CD44v6, CD44v4–10, or CD44v8–10, accounting for the diverse heterogeneity of the CD44 family [31, 44] (Fig. 6.2).

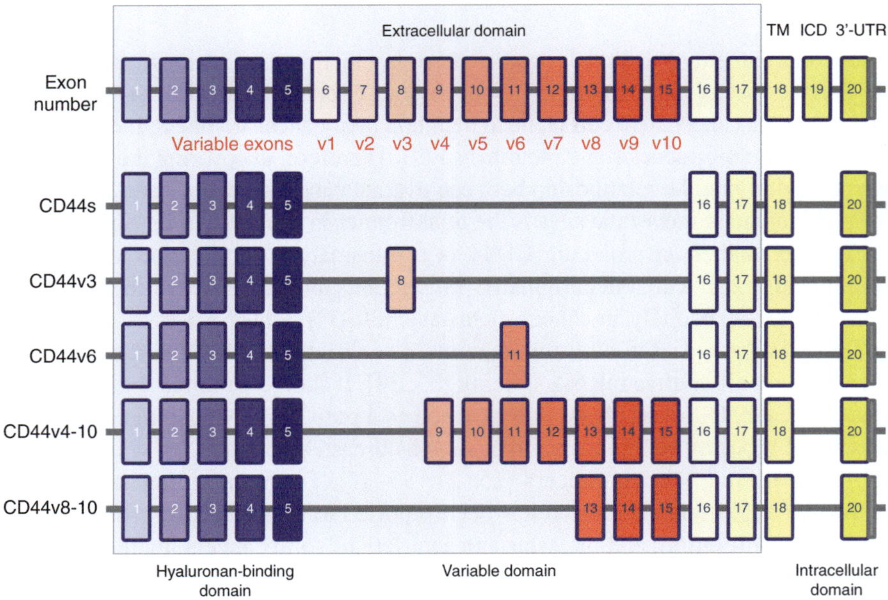

Fig. 6.2 CD44 genomic structure. The CD44 gene contains 20 exons, 10 of which can be regulated by alternative splicing of variant exons. The smallest CD44 standard includes exons 1–5 and 16–20. The CD44 variant is generated by the insertion of variant exons (exons 6–15). The structures of the four most common CD44 variant isoforms, CD44v3, CD44v6, CD44v4–10, and CD44v8–10, are shown. Abbreviations: TM, transmembrane region; ICD, intracellular region; UTR, untranslated region

The CD44 protein mainly consists of three functional domains, including the extracellular domain, the transmembrane domain, and the intracellular domain [44] (Fig. 6.3). These primary domains are responsible for the complexity of CD44 signaling [48]. The extracellular domain senses and interacts with the external microenvironment [49]. This is the most diverse region of the CD44 molecule, containing four regions designated as the hyaluronan-binding domain, the link domain, the basic motif, and the stem structure [50]. Among them, the stem structure can be enlarged by the alternatively spliced variant exons (v1–v10) of CD44 [51]. The transmembrane domain provides a platform for CD44 oligomer formation and an avenue for interacting with cofactors and adaptor proteins [49]. The CD44 intracellular domain is essential for the function of CD44 in signaling transduction and regulates the target genes, including cyclin D1, MMP-9, c-myc, and Twist [42, 44, 52] (Fig. 6.3).

CD44s is broadly expressed in various types of normal and malignant cells, including lymphocytes, macrophages, granulocytes, erythrocytes, epithelial cells, and some epithelial cancer cells [53]. On the other hand, a number of CD44 variant isoforms have tissue-specific expression and have been extensively evaluated as specific cell surface markers of malignant cells [53, 54]. In the context of these facts, further fundamental studies are needed in order to elucidate the functional

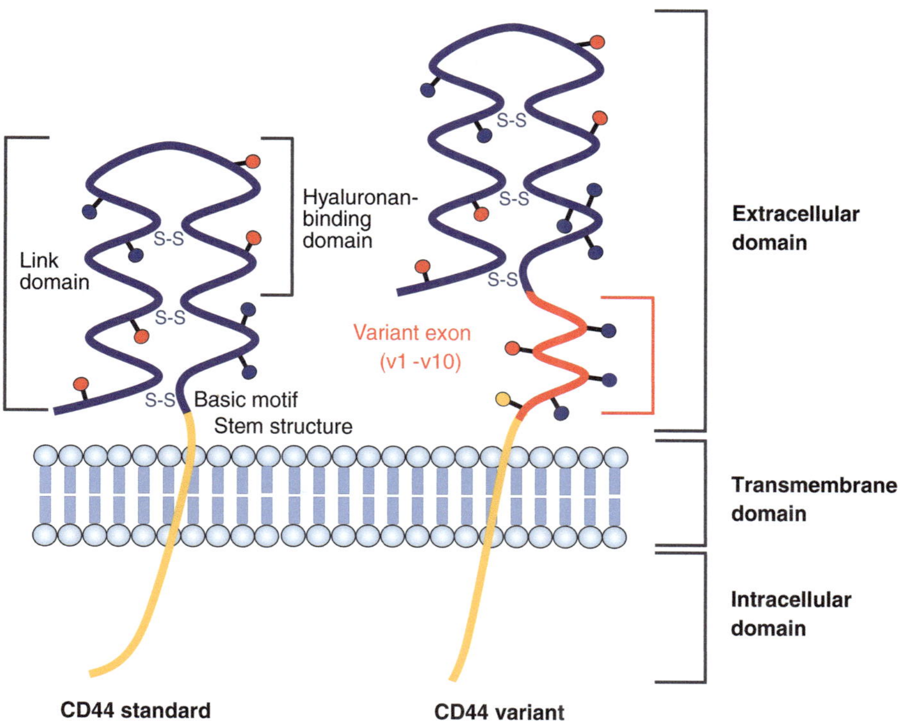

Fig. 6.3 CD44 protein structure. The CD44 transmembrane glycoprotein is composed of the extracellular domain, transmembrane domain, and intracellular domain. The stem structure can be enlarged through alternatively spliced variant exons (v1–v10) of CD44

role of CD44 and its variant isoforms in regulating the behavior of normal and malignant cells, respectively.

6.3 CD44 and Cancer Stemness

The cancer stem cell model postulates that cancer stem cells contribute to sustaining carcinogenesis and establishing the phenotypic and functional heterogeneity in a hierarchical organization of tumors [55, 56]. Recently, this model has received wide attention because it provides an explanation for tumor metastasis, resistance to conventional therapies, and eventual tumor recurrence [14, 21, 57]. A number of research concerning the expression of specific stemness markers of cancer cells have identified CD44 as one of the major cell surface markers associated with cancer stem cells in several types of epithelial malignancies [15, 31] (Table 6.1).

Regarding the relationship between CD44 and ovarian cancer stem cells, in 2008, Zhang et al. identified and characterized a self-renewing subpopulation of ovarian cancer stem cells that possess the ability to serially propagate their original tumor phenotype in a mouse model [58]. The investigators showed that sphere-forming cells coexpress CD44 and stem cell factor receptor CD117, and those distinct markers, which are used to isolate highly tumorigenic ovarian cancer stem cells, contribute to enhanced chemoresistance to paclitaxel and cisplatin. In addition, Alvero et al. revealed the molecular biological characterization of CD44-positive ovarian cancer stem cells [59]. CD44-positive cancer cells are inseparably connected with the expression of myeloid differentiation factor 88 (MyD88), a critical component of the Toll-like receptors (TLRs) in nuclear factor kappa B (NF-κB) signaling pathways, and promote a pro-inflammatory microenvironment, cytokine and chemokine production, and resistance to conventional chemotherapy in hierarchically organized ovarian tumors [59]. These findings suggested that CD44-positive ovarian cancer stem cells play a pivotal role in communicating with the specialized niche microenvironment that closely regulates the cancer stemness and governs the cellular fate of cancer stem cells [59].

Regarding the identification of a combination of cancer stem cell markers for ovarian cancer, previous reports demonstrated that aldehyde dehydrogenase

Table 6.1 Cancer stem cell markers in distinct types of solid cancers

Cancer type	Cancer stem cell marker
Breast	CD44 CD90 CD133 ALDH1 CXCR4 EpCAM
Colon	CD44 ABCB5 ALDH1 CD24 CD26 CD133 EpCAM LGR5
Giloma	α_6-integrin CD15 CD90 Nestin
Liver	CD44 CD13 CD24 CD90 CD133 EpCAM
Lung	ABCG2 ALDH1 CD90 CD117 CD133 CXCR4
Melanoma	ABCB5 ALDH1 CD20 CD133 CD271
Pancreatic	CD44 ABCG2 ALDH1 CD24 CD133 CXCR4 Nestin
Prostate	CD44 $\alpha_2\beta_1$-integrin α_6-integrin ALDH1 CD133 CD166
Ovarian	CD44 ALDH1 CD24 CD117 CD133 EpCAM

isoform 1 (ALDH1)-expressing cells isolated from ovarian cancer cell lines possess cancer stem cell properties and are enriched in cell surface marker CD44, indicating that the combination of ALDH1 and CD44 might define ovarian cancer stem cells more stringently [60, 61]. On another front, our research group previously identified a subpopulation of epithelial cell adhesion molecule (EpCAM)-expressing cells as candidates for ovarian cancer stem cells in an established mouse ovarian tumor model generated by transduction of defined genetic alterations [18]. We also found that EpCAM-expressing cancer cells manifest high expression of CD44, suggesting that these are common markers for ovarian cancer stem cells [18, 62]. Up to now, the candidates suggested to exhibit the qualities of ovarian cancer stem cells include CD44, CD117, ALDH1, CD133, CD24, or EpCAM [18, 58, 60, 63–65] (Table 6.1). However, it still remains challenging to identify one single cell surface marker or a combination of several markers to specifically identify all ovarian cancer stem cells [19, 56]. In fact, the currently available ovarian cancer stem cell markers look promising, but these markers are not always ubiquitous in a given tumor, and they might not be specific to each histological subtype of ovarian cancer [21]. In view of this, further studies toward a comprehensive understanding of ovarian cancer stem cell markers are required to consistently enrich the subpopulation of cancer stem cells.

6.4 CD44 Variant 6 and Cancer Metastasis

6.4.1 CD44 Variant 6 and Peritoneal Disseminated Metastasis

A milestone in cancer research on CD44 was the identification of CD44v6 as one of the definitive metastatic determinants in various types of solid cancers, including colon cancer, breast cancer, and melanoma [66–68]. However, the functional role of CD44v6 in ovarian cancer metastasis with respect to cancer stem cells has remained largely undetermined [69].

Ovarian cancer is fairly characterized by rapid growth, spread of multiple peritoneal disseminated tumors, and the accumulation of ascites [1, 70]. The biological behavior of peritoneal metastasis of ovarian cancer is distinctive, differing markedly from the widely known mechanisms of hematogenous metastasis occurring in most other solid cancers [71, 72]. For instance, in certain cancers, such as breast, colon, or lung cancer, cancer cells frequently go through a series of steps, such as local invasion, intravasation, travel through the circulation system, extravasation, and metastatic colonization, before they metastasize to other parenchymal organs, such as the brain, liver, or bone [73, 74]. On the other hand, fundamentally, peritoneal disseminated metastasis appears to be easier for some ovarian cancer cells [70, 75]. Once some specific ovarian cancer cells have detached the primary tumors as single cells or as spheroid clusters, they are able to metastasize through a direct mechanism to the peritoneum [70, 76].

In view of this background, we have previously evaluated whether a subpopulation of CD44v6-positive ovarian cancer cells play a functional role in peritoneal disseminated metastasis [20]. Clinicopathological analysis of CD44v6 expression revealed that peritoneal disseminated tumors contained a significantly higher percentage of CD44v6-positive cells than corresponding primary ovarian tumors, suggesting that CD44v6-positive cells are clinically correlated with peritoneal metastasis, and the pelvic peritoneum might have the potential to form part of the niche microenvironment associated with tumor initiation, metastatic colonization, and the formation of disseminated tumors in the peritoneal cavity [20, 28, 77]. Furthermore, consistent with clinical observations, in an in vivo mouse model, FACS-sorted CD44v6-positive cancer cells generated extensive peritoneal disseminated tumors, resulting in massive abdominal distension by hemorrhagic ascites, whereas FACS-sorted CD44v6-negative cells showed little capability to form disseminated tumors in the peritoneal cavity. Moreover, it was noteworthy that limiting dilution assay demonstrated that CD44v6-positive cancer cells had a greater tumor-initiating ability than CD44v6-negative cells, indicating that the subpopulation of CD44v6-positive cells is highly efficient at peritoneal metastatic dissemination and has the potential to serve as specialized metastasis-initiating cells [16, 20] (Fig. 6.4).

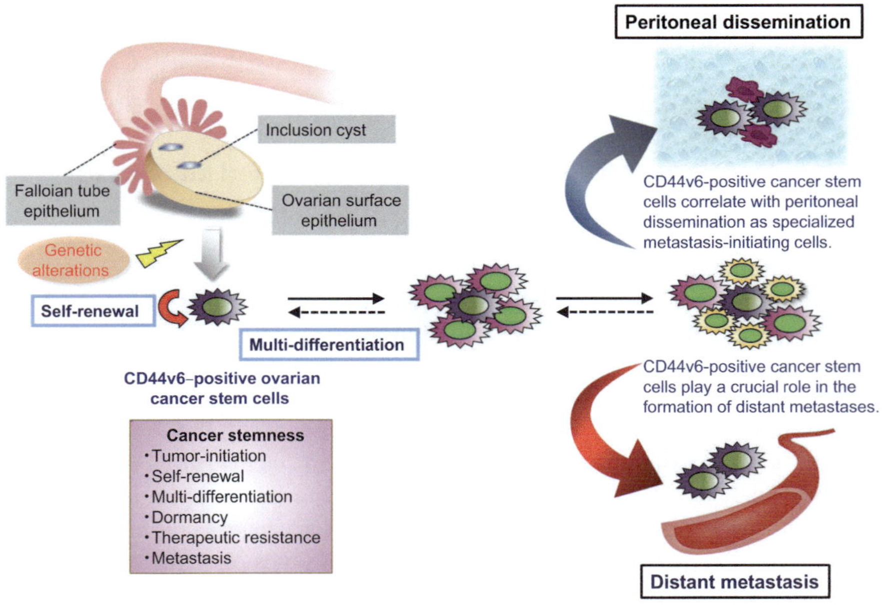

Fig. 6.4 Schema of ovarian cancer stem cell model for tumor initiation, development, and metastasis. Cancer stem cells have been regarded to originate from genetic alterations in normal somatic stem cells that deregulate their physiological programs; however, the cells of origin for ovarian cancers remain to be elucidated. Ovarian cancer stem cells possess the ability to self-renew and differentiate into various types of cancer cells and inherently contribute to tumor metastasis, therapy resistance, and recurrence in metastatic sites. CD44v6-positive ovarian cancer stem cells are principally able to metastasize to the peritoneum as multiple disseminated tumors or distant parenchymal organs as specialized metastasis-initiating cells. In addition, CD44v6 represents a predictive biomarker for cancer metastasis and an unfavorable prognosis in patients with ovarian cancer

6.4.2 CD44 Variant 6 and Distant Parenchymal Metastasis

As a general rule, intraperitoneal dissemination is considered the most common route by which ovarian cancer cells spread; however, ovarian cancer can metastasize through blood vessels to the parenchyma of distant organs or in parallel through the lymphatic nodes [70]. Because the majority of ovarian cancer-related deaths are directly attributed to the development of metastatic disease [78], recognition of the molecular mechanisms underlying the development of distant metastases provides a new opportunity for better management and targeted therapeutic interventions in ovarian cancer patients.

Our research group has recently demonstrated for the first time that CD44v6 represents a predictive biomarker for distant metastasis and CD44v6-positive ovarian cancer cells play a central role in the development of distant metastases in parenchymal organs [79]. Intriguingly, increased CD44v6 expression was significantly associated with distant metastasis at the time of ovarian cancer diagnosis, whereas the expression of CD44v6 was not involved in the presence or absence of lymph node metastases. Furthermore, Kaplan–Meier analysis revealed that distant metastasis-free survival significantly differed between the CD44v6-high and CD44v6-low groups in patients with stage I–III ovarian cancer, and more importantly, multivariate analysis demonstrated that CD44v6 expression is an independent risk factor for distant metastatic recurrence, suggesting that CD44v6 expression in primary tumors is correlated with increased distant metastasis and CD44v6 is a potential predictive biomarker of the development of distant metastatic recurrence [79]. Taken together, these findings suggest that CD44v6 represents a promising molecular therapeutic target for eradicating ovarian cancer stem cells and metastasis-initiating cells as the root of ovarian cancer [28] (Fig. 6.4).

6.4.3 CD44 Variant 6 and the Epithelial–Mesenchymal Transition

The epithelial–mesenchymal transition (EMT) is a highly conserved cellular program that enables polarized immotile epithelial cells to convert to motile mesenchymal cells [80–82]. This principal phenomenon was initially observed during several critical stages of embryonic development and has more recently been correlated with acceleration of cancer invasion and metastasis [82, 83]. Recent evidence indicated that cancer stem cells contribute to the initiation of tumor metastasis, a notion strengthened by the correlation between cancer stem cells and EMT. Initial evidence by Robert Weinberg's group demonstrated that EMT by ectopic expression of the transcription factors Twist or Snail was involved in cancer stem cell qualities and stem cell marker expression [84]. In this regard, a large body of experimental research by many investigators has evaluated the close relationship between CD44 and the EMT phenomenon [31, 85–87].

In patients with ovarian cancer, when the cancer cells start their metastatic journey from primary tumors, it is known that they mainly undergo an EMT in order to facilitate cancer metastasis [70, 88]. However, the contribution of CD44v6 to the EMT process in ovarian cancer has remained unresolved. We previously revealed the important role of the subpopulation of CD44v6-positive cells in the EMT of ovarian cancer metastasis [20]. The ability of CD44v6-positive cancer cells to undergo EMT was investigated by determining the expression of EMT-related proteins in FACS-sorted CD44v6-positive cells versus CD44v6-negative cells. As a result, the expression of E-cadherin in CD44v6-positive cells was lower than that in CD44v6-negative cells, whereas the expression of N-cadherin, fibronectin, and vimentin in CD44v6-positive cells was higher than in CD44v6-negative cells, suggesting that CD44v6 expression is involved in the mesenchymal phenotype, and the subpopulation of CD44v6-positive cells regulates the metastatic capability of ovarian cancer cells, which is relevant to the process of EMT [20].

6.4.4 CD44 Variant 6 and the Metastatic Niche

The organ-specific pattern of metastatic spread has been known for centuries. However, uncovering the explanation for cancer metastasis to specific organs has been challenging ever since the "seed and soil" hypothesis was postulated by Stephen Paget more than a century ago [12]. In recent history, growing evidence has suggested that the primary tumors are able to drive metastasis by leading to the formation of a supportive tumor microenvironment in secondary or distant metastatic sites, termed the "pre-metastatic niche" [26, 72, 89]. In particular, previous studies demonstrated the functional role of CD44v6 in regulating a tumor microenvironment that has a significant impact on pre-metastatic niche formation, which depends on the cooperative activity of the CD44v6-dependent organization of tumor-derived soluble fractions and exosomes [49, 77]. Regarding niche formation for metastatic dissemination of ovarian cancer, an earlier study showed that peritoneal mesothelial cells have the potential to form the niche microenvironment, which enhances the ovarian cancer stemness [20, 28]. In the future, further studies are required in order to establish the causal relationship between CD44v6 and the formation of a pre-metastatic niche in peritoneal or distant metastatic sites in ovarian cancer patients.

Taken together, the apparent association between the expression of CD44v6 and ovarian cancer metastasis raises the possibility that a subpopulation of cancer stem cells actually forms the "seeds," while at the same time, the possible involvement of CD44v6 in creating the metastatic niche microenvironment suggests that the "soil" is the specialized cancer stem cell niche [29, 89]. Given that CD44v6-positive cancer stem cells are responsible for the development of clinically significant ovarian cancer metastasis, definitive therapeutic strategies should target the CD44v6-positive cell population and the related niche microenvironment.

6.5 CD44 Variant 6 and Clinical Outcome

The dynamic and functional properties of cancer stem cells provide a possible explanation for clinical observations, such as tumor metastasis, therapy resistance, and disease relapse [72, 79]. If cancer stem cells are actually involved in these clinical problems, their functional characteristics should correlate with clinical outcomes in cancer patients. In fact, the distinct expression of cancer stem cell markers has been linked to survival, and a growing number of studies have shown a close relationship between the presence of cancer stem cells and worse clinical outcomes in different types of solid cancers [10, 90–92].

Although some previous studies focused on the potential correlation of CD44 and its variant isoforms with ovarian cancer survival to address the diagnostic and prognostic values of this cell surface marker, there has been no consensus concerning this issue [69, 93–95]. In order to address these unresolved questions, we evaluated the association between CD44v6 expression and aspects of ovarian cancer survival based on immunohistochemical analysis. As the result, in the evaluation of the sites of primary ovarian tumors, we found that the tumors containing at least 10% CD44v6-positive cancer cells showed significantly shortened overall survival compared to those containing less than 10% CD44v6-positive cells [20, 79]. In addition, it should be noted that a multivariate Cox proportional hazards model demonstrated that CD44v6 expression is an independent prognostic factor for the overall survival of patients with advanced ovarian cancer [20].

Collectively, these findings support the premise that the expression of CD44v6 can be used as a predictive and prognostic biomarker in a clinical setting of ovarian cancer patients. In order to achieve the ultimate goal of developing clinical measures that will improve the clinical outcomes of ovarian cancer patients, CD44v6-targeted strategies should make a promising therapeutic avenue for ovarian cancer.

Conclusions

The encompassing goal of cancer stem cell research is to develop our comprehension of the molecular basis of cancer in the correlation between cancer stem cells and cancer stem cell niche microenvironment and guide clinical decision-making in association with the proper management of cancer patients, biomarker-based clinical trials, and the development of novel therapeutic interventions. Up to now, comprehensive cancer stem cell research has provided a strong foundation for clarifying the dynamic molecular mechanisms of cancer biology and stimulated a diversity of gynecological cancer research [96, 97]. Truly, the increased enthusiasm for ovarian cancer stem cell research has led to the development of promising disease biomarkers and novel molecular-targeted therapies [10, 96, 98]. Given that CD44v6 has been identified as key players in regulating ovarian cancer metastasis and in creating their niche microenvironment, CD44v6 seems to be a promising target for specific therapy directed against cancer stem cells and cancer stem cell niche. In order to translate these scientific findings into meaningful and practicable clinical end points, further elucidation of the causal relationship between ovarian cancer stem cells and cancer stem cell niche in metastatic sites is required.

We hope that ovarian cancer treatments directed toward the eradication of the subpopulation of CD44v6-positive metastatic cancer stem cells will lead to higher survival rates and brighter prognoses for patients with this life-threatening malignancy. Although we still have a long way to go, we are also convinced that the innovative journey to uncover answers and extirpate ovarian cancer stem cells is sure to be an exciting and great one.

References

1. Bast RC Jr, Hennessy B, Mills GB. The biology of ovarian cancer: new opportunities for translation. Nat Rev Cancer. 2009;9:415–28.
2. Agarwal R, Kaye SB. Ovarian cancer: strategies for overcoming resistance to chemotherapy. Nat Rev Cancer. 2003;3:502–16.
3. Heintz AP, Odicino F, Maisonneuve P, et al. Carcinoma of the ovary. Int J Gynaecol Obstet. 2003;83:135–66.
4. Buys SS, Partridge E, Black A, et al. Effect of screening on ovarian cancer mortality: the prostate, lung, colorectal and ovarian (PLCO) cancer screening randomized controlled trial. JAMA. 2011;305:2295–303.
5. Esselen KM, Cronin AM, Bixel K, et al. Use of CA-125 tests and computed tomographic scans for surveillance in ovarian cancer. JAMA Oncol. 2016;2:1427–33.
6. Katabuchi H, Okamura H. Cell biology of human ovarian surface epithelial cells and ovarian carcinogenesis. Med Electron Microsc. 2003;36:74–86.
7. Yap TA, Carden CP, Kaye SB. Beyond chemotherapy: targeted therapies in ovarian cancer. Nat Rev Cancer. 2009;9:167–81.
8. Hennessy BT, Coleman RL, Markman M. Ovarian cancer. Lancet. 2009;374:1371–82.
9. Meyer LA, Cronin AM, Sun CC, et al. Use and effectiveness of neoadjuvant chemotherapy for treatment of ovarian cancer. J Clin Oncol. 2016;34:3854–63.
10. Tayama S, Motohara T, Narantuya D, et al. The impact of EpCAM expression on response to chemotherapy and clinical outcomes in patients with epithelial ovarian cancer. Oncotarget. 2017;8:44312–25.
11. Miranda F, Mannion D, Liu S, et al. Salt-inducible kinase 2 couples ovarian cancer cell metabolism with survival at the adipocyte-rich metastatic niche. Cancer Cell. 2016;30:273–89.
12. Paget S. The distribution of secondary growths in cancer of the breast. 1889. Cancer Metastasis Rev. 1989;8:98–101.
13. Fidler IJ. The pathogenesis of cancer metastasis: the 'seed and soil' hypothesis revisited. Nat Rev Cancer. 2003;3:453–8.
14. Reya T, Morrison SJ, Clarke MF, Weissman IL. Stem cells, cancer, and cancer stem cells. Nature. 2001;414:105–11.
15. Visvader JE, Lindeman GJ. Cancer stem cells in solid tumours: accumulating evidence and unresolved questions. Nat Rev Cancer. 2008;8:755–68.
16. Baccelli I, Trumpp A. The evolving concept of cancer and metastasis stem cells. J Cell Biol. 2012;198:281–93.
17. Bapat SA, Mali AM, Koppikar CB, Kurrey NK. Stem and progenitor-like cells contribute to the aggressive behavior of human epithelial ovarian cancer. Cancer Res. 2005;65:3025–9.
18. Motohara T, Masuko S, Ishimoto T, et al. Transient depletion of p53 followed by transduction of c-Myc and K-Ras converts ovarian stem-like cells into tumor-initiating cells. Carcinogenesis. 2011;32:1597–606.
19. Ahmed N, Abubaker K, Findlay JK. Ovarian cancer stem cells: molecular concepts and relevance as therapeutic targets. Mol Asp Med. 2014;39:110–25.

20. Tjhay F, Motohara T, Tayama S, et al. CD44 variant 6 is correlated with peritoneal dissemination and poor prognosis in patients with advanced epithelial ovarian cancer. Cancer Sci. 2015;106:1421–8.
21. Visvader JE, Lindeman GJ. Cancer stem cells: current status and evolving complexities. Cell Stem Cell. 2012;10:717–28.
22. Medema JP. Cancer stem cells: the challenges ahead. Nat Cell Biol. 2013;15:338–44.
23. Schofield R. The relationship between the spleen colony-forming cell and the haemopoietic stem cell. Blood Cells. 1978;4:7–25.
24. Borovski T, De Sousa EMF, Vermeulen L, Medema JP. Cancer stem cell niche: the place to be. Cancer Res. 2011;71:634–9.
25. Calabrese C, Poppleton H, Kocak M, et al. A perivascular niche for brain tumor stem cells. Cancer Cell. 2007;11:69–82.
26. Kaplan RN, Rafii S, Lyden D. Preparing the "soil": the premetastatic niche. Cancer Res. 2006;66:11089–93.
27. Erler JT, Bennewith KL, Cox TR, et al. Hypoxia-induced lysyl oxidase is a critical mediator of bone marrow cell recruitment to form the premetastatic niche. Cancer Cell. 2009;15:35–44.
28. Mitsui H, Shibata K, Suzuki S, et al. Functional interaction between peritoneal mesothelial cells and stem cells of ovarian yolk sac tumor (SC-OYST) in peritoneal dissemination. Gynecol Oncol. 2012;124:303–10.
29. Wicha MS. Cancer stem cells and metastasis: lethal seeds. Clin Cancer Res. 2006;12:5606–7.
30. Orian-Rousseau V. CD44, a therapeutic target for metastasising tumours. Eur J Cancer. 2010;46:1271–7.
31. Zoller M. CD44: can a cancer-initiating cell profit from an abundantly expressed molecule? Nat Rev Cancer. 2011;11:254–67.
32. Spring FA, Dalchau R, Daniels GL, et al. The Ina and Inb blood group antigens are located on a glycoprotein of 80,000 MW (the CDw44 glycoprotein) whose expression is influenced by the In(Lu) gene. Immunology. 1988;64:37–43.
33. Rodrigo JP, Dominguez F, Alvarez C, Gonzalez MV, Herrero A, Suarez C. Clinicopathologic significance of expression of CD44s and CD44v6 isoforms in squamous cell carcinoma of the supraglottic larynx. Am J Clin Pathol. 2002;118:67–72.
34. Erb U, Megaptche AP, Gu X, Buchler MW, Zoller M. CD44 standard and CD44v10 isoform expression on leukemia cells distinctly influences niche embedding of hematopoietic stem cells. J Hematol Oncol. 2014;7:29.
35. Al-Hajj M, Wicha MS, Benito-Hernandez A, Morrison SJ, Clarke MF. Prospective identification of tumorigenic breast cancer cells. Proc Natl Acad Sci U S A. 2003;100:3983–8.
36. Ishimoto T, Nagano O, Yae T, et al. CD44 variant regulates redox status in cancer cells by stabilizing the xCT subunit of system xc(−) and thereby promotes tumor growth. Cancer Cell. 2011;19:387–400.
37. Yae T, Tsuchihashi K, Ishimoto T, et al. Alternative splicing of CD44 mRNA by ESRP1 enhances lung colonization of metastatic cancer cell. Nat Commun. 2012;3:883.
38. Yamane N, Tsujitani S, Makino M, Maeta M, Kaibara N. Soluble CD44 variant 6 as a prognostic indicator in patients with colorectal cancer. Oncology. 1999;56:232–8.
39. Jijiwa M, Demir H, Gupta S, et al. CD44v6 regulates growth of brain tumor stem cells partially through the AKT-mediated pathway. PLoS One. 2011;6:e24217.
40. Shi J, Zhou Z, Di W, Li N. Correlation of CD44v6 expression with ovarian cancer progression and recurrence. BMC Cancer. 2013;13:182.
41. Todaro M, Gaggianesi M, Catalano V, et al. CD44v6 is a marker of constitutive and reprogrammed cancer stem cells driving colon cancer metastasis. Cell Stem Cell. 2014;14:342–56.
42. Nagano O, Saya H. Mechanism and biological significance of CD44 cleavage. Cancer Sci. 2004;95:930–5.
43. Naor D, Sionov RV, Ish-Shalom D. CD44: structure, function, and association with the malignant process. Adv Cancer Res. 1997;71:241–319.
44. Ponta H, Sherman L, Herrlich PA. CD44: from adhesion molecules to signalling regulators. Nat Rev Mol Cell Biol. 2003;4:33–45.

45. Senbanjo LT, Chellaiah MA. CD44: a multifunctional cell surface adhesion receptor is a regulator of progression and metastasis of cancer cells. Front Cell Dev Biol. 2017;5:18.
46. Yan Y, Zuo X, Wei D. Concise review: emerging role of CD44 in cancer stem cells: a promising biomarker and therapeutic target. Stem Cells Transl Med. 2015;4:1033–43.
47. Misra S, Heldin P, Hascall VC, et al. Hyaluronan-CD44 interactions as potential targets for cancer therapy. FEBS J. 2011;278:1429–43.
48. Goldstein LA, Zhou DF, Picker LJ, et al. A human lymphocyte homing receptor, the hermes antigen, is related to cartilage proteoglycan core and link proteins. Cell. 1989;56:1063–72.
49. Williams K, Motiani K, Giridhar PV, Kasper S. CD44 integrates signaling in normal stem cell, cancer stem cell and (pre)metastatic niches. Exp Biol Med (Maywood). 2013;238:324–38.
50. Underhill C. CD44: the hyaluronan receptor. J Cell Sci. 1992;103:293–8.
51. Brown RL, Reinke LM, Damerow MS, et al. CD44 splice isoform switching in human and mouse epithelium is essential for epithelial-mesenchymal transition and breast cancer progression. J Clin Invest. 2011;121:1064–74.
52. Okamoto I, Kawano Y, Murakami D, et al. Proteolytic release of CD44 intracellular domain and its role in the CD44 signaling pathway. J Cell Biol. 2001;155:755–62.
53. Nagano O, Okazaki S, Saya H. Redox regulation in stem-like cancer cells by CD44 variant isoforms. Oncogene. 2013;32:5191–8.
54. Orian-Rousseau V, Ponta H. Perspectives of CD44 targeting therapies. Arch Toxicol. 2015;89:3–14.
55. Lupia M, Cavallaro U. Ovarian cancer stem cells: still an elusive entity? Mol Cancer. 2017;16:64.
56. Garson K, Vanderhyden BC. Epithelial ovarian cancer stem cells: underlying complexity of a simple paradigm. Reproduction. 2015;149:R59–70.
57. Ahmed N, Abubaker K, Findlay J, Quinn M. Cancerous ovarian stem cells: obscure targets for therapy but relevant to chemoresistance. J Cell Biochem. 2013;114:21–34.
58. Zhang S, Balch C, Chan MW, et al. Identification and characterization of ovarian cancer-initiating cells from primary human tumors. Cancer Res. 2008;68:4311–20.
59. Alvero AB, Chen R, Fu HH, et al. Molecular phenotyping of human ovarian cancer stem cells unravels the mechanisms for repair and chemoresistance. Cell Cycle. 2009;8:158–66.
60. Wang YC, Yo YT, Lee HY, et al. ALDH1-bright epithelial ovarian cancer cells are associated with CD44 expression, drug resistance, and poor clinical outcome. Am J Pathol. 2012;180:1159–69.
61. Charafe-Jauffret E, Ginestier C, Iovino F, et al. Aldehyde dehydrogenase 1-positive cancer stem cells mediate metastasis and poor clinical outcome in inflammatory breast cancer. Clin Cancer Res. 2010;16:45–55.
62. Marhaba R, Klingbeil P, Nuebel T, Nazarenko I, Buechler MW, Zoeller M. CD44 and EpCAM: cancer-initiating cell markers. Curr Mol Med. 2008;8:784–804.
63. Kuroda T, Hirohashi Y, Torigoe T, et al. ALDH1-high ovarian cancer stem-like cells can be isolated from serous and clear cell adenocarcinoma cells, and ALDH1 high expression is associated with poor prognosis. PLoS One. 2013;8:e65158.
64. Baba T, Convery PA, Matsumura N, et al. Epigenetic regulation of CD133 and tumorigenicity of CD133+ ovarian cancer cells. Oncogene. 2009;28:209–18.
65. Gao MQ, Choi YP, Kang S, Youn JH, Cho NH. CD24+ cells from hierarchically organized ovarian cancer are enriched in cancer stem cells. Oncogene. 2010;29:2672–80.
66. Gunthert U, Hofmann M, Rudy W, et al. A new variant of glycoprotein CD44 confers metastatic potential to rat carcinoma cells. Cell. 1991;65:13–24.
67. Hofmann M, Rudy W, Zoller M, et al. CD44 splice variants confer metastatic behavior in rats: homologous sequences are expressed in human tumor cell lines. Cancer Res. 1991;51:5292–7.
68. Tanabe KK, Nishi T, Saya H. Novel variants of CD44 arising from alternative splicing: changes in the CD44 alternative splicing pattern of MCF-7 breast carcinoma cells treated with hyaluronidase. Mol Carcinog. 1993;7:212–20.
69. Hong SC, Song JY, Lee JK, et al. Significance of CD44v6 expression in gynecologic malignancies. J Obstet Gynaecol Res. 2006;32:379–86.

70. Lengyel E. Ovarian cancer development and metastasis. Am J Pathol. 2010;177:1053–64.
71. Brabletz T, Hlubek F, Spaderna S, et al. Invasion and metastasis in colorectal cancer: epithelial-mesenchymal transition, mesenchymal-epithelial transition, stem cells and beta-catenin. Cells Tissues Organs. 2005;179:56–65.
72. Oskarsson T, Batlle E, Massague J. Metastatic stem cells: sources, niches, and vital pathways. Cell Stem Cell. 2014;14:306–21.
73. Warren BA, Shubik P, Feldman R. Metastasis via the blood stream: the method of intravasation of tumor cells in a transplantable melanoma of the hamster. Cancer Lett. 1978;4:245–51.
74. Schwartz RS, Erban JK. Timing of metastasis in breast cancer. N Engl J Med. 2017;376:2486–8.
75. Gardner MJ, Catterall JB, Jones LM, Turner GA. Human ovarian tumour cells can bind hyaluronic acid via membrane CD44: a possible step in peritoneal metastasis. Clin Exp Metastasis. 1996;14:325–34.
76. Ahmed N, Stenvers KL. Getting to know ovarian cancer ascites: opportunities for targeted therapy-based translational research. Front Oncol. 2013;3:256.
77. Jung T, Castellana D, Klingbeil P, et al. CD44v6 dependence of premetastatic niche preparation by exosomes. Neoplasia. 2009;11:1093–105.
78. Guth U, Huang DJ, Bauer G, Stieger M, Wight E, Singer G. Metastatic patterns at autopsy in patients with ovarian carcinoma. Cancer. 2007;110:1272–80.
79. Motohara T, Fujimoto K, Tayama S, et al. CD44 variant 6 as a predictive biomarker for distant metastasis in patients with epithelial ovarian cancer. Obstet Gynecol. 2016;127:1003–11.
80. Yang J, Weinberg RA. Epithelial-mesenchymal transition: at the crossroads of development and tumor metastasis. Dev Cell. 2008;14:818–29.
81. Hay ED. An overview of epithelio-mesenchymal transformation. Acta Anat. 1995;154:8–20.
82. Cheng GZ, Chan J, Wang Q, Zhang W, Sun CD, Wang LH. Twist transcriptionally up-regulates AKT2 in breast cancer cells leading to increased migration, invasion, and resistance to paclitaxel. Cancer Res. 2007;67:1979–87.
83. Radisky DC. Epithelial-mesenchymal transition. J Cell Sci. 2005;118:4325–6.
84. Mani SA, Guo W, Liao MJ, et al. The epithelial-mesenchymal transition generates cells with properties of stem cells. Cell. 2008;133:704–15.
85. Xu H, Tian Y, Yuan X, et al. The role of CD44 in epithelial-mesenchymal transition and cancer development. Onco Targets Ther. 2015;8:3783–92.
86. Zhang Y, Wei J, Wang H, et al. Epithelial mesenchymal transition correlates with CD24+CD44+ and CD133+ cells in pancreatic cancer. Oncol Rep. 2012;27:1599–605.
87. Blick T, Hugo H, Widodo E, et al. Epithelial mesenchymal transition traits in human breast cancer cell lines parallel the CD44(hi/)CD24 (lo/−) stem cell phenotype in human breast cancer. J Mammary Gland Biol Neoplasia. 2010;15:235–52.
88. Veatch AL, Carson LF, Ramakrishnan S. Differential expression of the cell-cell adhesion molecule E-cadherin in ascites and solid human ovarian tumor cells. Int J Cancer. 1994;58:393–9.
89. Liu Y, Cao X. Characteristics and significance of the pre-metastatic niche. Cancer Cell. 2016;30:668–81.
90. Coco C, Zannoni GF, Caredda E, et al. Increased expression of CD133 and reduced dystroglycan expression are strong predictors of poor outcome in colon cancer patients. J Exp Clin Cancer Res. 2012;31:71.
91. Langan RC, Mullinax JE, Raiji MT, et al. Colorectal cancer biomarkers and the potential role of cancer stem cells. J Cancer. 2013;4:241–50.
92. Piscuoglio S, Lehmann FS, Zlobec I, et al. Effect of EpCAM, CD44, CD133 and CD166 expression on patient survival in tumours of the ampulla of Vater. J Clin Pathol. 2012;65:140–5.
93. Sliutz G, Tempfer C, Winkler S, Kohlberger P, Reinthaller A, Kainz C. Immunohistochemical and serological evaluation of CD44 splice variants in human ovarian cancer. Br J Cancer. 1995;72:1494–7.
94. Sakai K, Kaku T, Kamura T, et al. Comparison of p53, Ki-67, and CD44v6 expression between primary and matched metastatic lesions in ovarian cancer. Gynecol Oncol. 1999;72:360–6.
95. Lin J, Ding D. The prognostic role of the cancer stem cell marker CD44 in ovarian cancer: a meta-analysis. Cancer Cell Int. 2017;17:8.

96. Liu J, Matulonis UA. New strategies in ovarian cancer: translating the molecular complexity of ovarian cancer into treatment advances. Clin Cancer Res. 2014;20:5150–6.
97. Patch AM, Christie EL, Etemadmoghadam D, et al. Whole-genome characterization of chemo-resistant ovarian cancer. Nature. 2015;521:489–94.
98. Petrillo M, Nero C, Amadio G, Gallo D, Fagotti A, Scambia G. Targeting the hallmarks of ovarian cancer: the big picture. Gynecol Oncol. 2016;142:176–83.

Clinical Management of Primary Ovarian Insufficiency

7

Yoshinori Okamura

Abstract

Primary ovarian insufficiency (POI) is typically used to describe ovarian dysfunction when the underlying mechanism is unknown, although genetic defects, chemotherapy, radiotherapy, or surgery is also known to be causative. The incidence is rare, affecting approximately 1 in 100 women under the age of 40 years. POI is characterized by a triad of amenorrhea, low serum estradiol levels, and elevated serum follicle-stimulating hormone levels. The condition leads not only to infertility but also to ovarian hormone deficiency that is associated with health risks that include menopausal symptoms, decreased bone mineral density, and increased cardiovascular risk. Hormone replacement therapy should therefore be commenced as soon as possible to prevent adverse health risks. In cancer patients who are to receive chemotherapy or radiotherapy, fertility preservation options should be offered, though embryo donation and adoption are considered as alternatives. Given that recent studies in genetically modified animal models have revealed that the underlying mechanism involves increased follicle activation and subsequent early follicle exhaustion, new therapeutic strategies are needed.

Keywords

Fertility preservation · Hormone replacement therapy · Follicle activation · Ovarian tissue cryopreservation · Gonadotropin-releasing hormone analog

Y. Okamura
Department of Obstetrics and Gynecology, Faculty of Life Sciences, Kumamoto University, Kumamoto, Japan

Department of Gynecology, Kumamoto General Hospital, Yatsushiro, Kumamoto, Japan
e-mail: yokamura@kuh.kumamoto-u.ac.jp

H. Katabuchi et al. (eds.), *Cell Biology of the Ovary*,
https://doi.org/10.1007/978-981-10-7941-2_7

7.1 Introduction

Primary ovarian insufficiency (POI) describes a broad range of impaired ovarian functions. Other terms have been used to define these unusual conditions, including premature ovarian failure, premature menopause, and hypergonadotropic hypogonadism [1, 2]. POI is defined as amenorrhea for at least 4 months, sex steroid deficiency, and elevations in two serum follicle-stimulating hormone (FSH) tests obtained at least 1 month apart (>40 IU/L) in a woman aged less than 40 years [3]. This condition is different from the menopause, as ovarian follicular activity might recover after diagnosis and even lead to pregnancy in some woman [4]. The median age of menopause is approximately 51 years, but the range is normally between 40 and 60 years [5]. Menopause before 40 years of age is considered nonphysiological.

7.2 Etiology

POI is a heterogeneous, multifactorial disorder, though the etiology is unknown in most cases (90%) [6]. Spontaneous POI affects approximately 1% of women before 40 years of age and approximately 0.1% of women before 30 years of age [3].

Many causes of POI can be classified as intraovarian causes or structural and numerical abnormalities of the X chromosome. Intraovarian causes include ovarian follicle dysfunction, such as signaling defects, enzyme deficiencies, and autoimmunity [7]. Structural and numerical abnormalities of the X chromosome include, among others, Turner's syndrome (45X) [8] and the fragile X mental retardation 1 (*FMR1*) gene mutation, which is responsible for the fragile X syndrome [9]. X chromosome abnormalities have been estimated to occur in 12%–14% of woman with POI [10, 11].

Turner's syndrome is characterized by X chromosome monosomy (45X) and occurs with an incidence of 1/2500 live births [8]. These patients present with primary amenorrhea due to rapid follicular degeneration. However, patients with mosaic Turner's syndrome (45X/46XX) have a milder phenotype and may present with secondary amenorrhea and hypergonadotropic hypogonadism.

The most frequent single-gene mutation associated with POI is the *FMR1* mutation [12]. This gene contains an unstable CGG trinucleotide repeat within the 5′ untranslated region, varying in length from 6 to 54 repeats in normal individuals, and which can expand during transmission to the next generation. Carriers of the pre-mutation have 55–199 CGG repeats, and it is estimated that approximately 23% of these women will develop POI [13]. Therefore, women with the pre-mutation allele have a substantially increased risk of POI [14].

Bone morphogenetic protein 15 (BMP15) is a member of the large family of transforming growth factor beta (TGF-β) proteins needed for embryo development and tissue formation. BMP15 is located at Xp 11.2 and is expressed exclusively in oocytes during folliculogenesis. Mutations of the BMP15 gene can cause POI [15].

Many other autosomal genes are involved in ovarian steroidogenesis and may cause POI [16] (Table 7.1).

Table 7.1 Genes involved in POI

Gene	Approved name	Chromosomal location
X chromosome		
BMP15	Bone morphogenic protein 15	Xp11.2
FMR1	Fragile X mental retardation 1	Xq27.3
FMR2	AF4/FMR2 family member 2	Xq28
Autosomal genes		
LHR (LHCGR)	Luteinizing hormone/choriogonadotropin receptor	2p16.3
FSHR	Follicle-stimulating hormone receptor	2p16.3
INHA	Inhibin alpha subunit	2q35
FOXL2	Forkhead box L2	3q22.3
FOXO3	Forkhead box O3	6q21
ESR1	Estrogen receptor 1	6q25.1–q25.2
SF1 (NR5A1)	Nuclear receptor subfamily 5 group A member 1	9q33.3
ESR2	Estrogen receptor 2	14q23.2–q23.3
CYP19A1	Cytochrome P450 family 19 subfamily A member 1	15q21.1

7.3 Animal Models

Loss of the pool of primordial follicles in the ovarian cortex is the underlying cause of POI. Primordial follicles are composed of oocytes surrounded by a layer of flat granulosa cells and are formed during fetal development. They then exist in a resting state until recruited for follicle growth, and although the mechanism of initiation of follicle growth is not fully understood, growth factors secreted from oocytes and granulosa cells are thought to be responsible for early follicle development. To preserve the pool of primordial follicles throughout reproductive life, they must be maintained in a dormant state. The molecular mechanism responsible for maintaining this dormancy and survival has been determined through studies using gene knockout models in mice.

Mice deficient in forkhead box O3 (Foxo3a) demonstrated premature activation of primordial follicles, with a resulting phenotype resembling that of POI [17]. This finding provided the first evidence that the phosphatidylinositol 3 kinase (PI3K) pathway was important in regulating follicle activation, with evidence that PI3K phosphorylation resulted in the activation of primordial follicles [18, 19]. The phosphatase and tensin homolog deleted on chromosome 10 (*PTEN*), which is a negative regulator of PI3K activity, can be deleted from the oocytes of primordial follicles in mice [18]. Subsequent research showed that this led to oocyte overgrowth and activation of the entire pool of primordial follicles, ultimately causing follicle depletion and POF [19].

Signaling mediated by PI3K converges primarily at 3-phosphoinositide-dependent kinase 1 (PDK1). To study the role of PI3K–PDK1 signaling in the oocytes of primordial follicles, the *PDK1* gene was deleted from the oocytes of primordial follicles. In mice lacking PDK1 in their oocytes, most primordial follicles were depleted from the dormant state around the onset of sexual maturity, causing POI during early adulthood [20]. Deletion of either PTEN or PDK1 in mouse oocytes resulted in POI but through different mechanisms. Deletion of PTEN results in overactivation of the primordial follicle pool, whereas deletion of PDK1 results in depletion of primordial follicles without activation. In double-knockout mice that lack both PTEN and PDK1 in oocytes, the overactivation of primordial follicles seen in PTEN-knockout mice was prevented [20]. This suggests that the overactivation noticed in PTEN-knockout mice was mediated primarily by PDK1 signaling.

Mammalian target of rapamycin (mTOR), a serine/threonine kinase that controls protein synthesis and cell growth, is regulated downstream by Akt activation. It exists in two biochemically and functionally different complexes, mTORC1 and mTORC2, according to its sensitivity to rapamycin [21]. S6 K1 is an important substrate of mTORC1 and can be phosphorylated by mTORC1 on its hydrophobic motif (T389) [22], which is a prerequisite for activation [23]. The tumor suppressor tuberous sclerosis complex 1 (Tsc1) negatively regulates mTORC1 [24]. In mutant mice with oocytes lacking the Tsc1 gene, the entire pool of primordial follicles was shown to be activated prematurely due to elevated mTORC1 activity, resulting in follicular depletion in early adulthood [25]. This indicates that mTORC1 is involved in primordial follicle activation and that the mTORC1–S6K1–rpS6 cascade might be important for the survival of primordial follicles, such as PI3K–PDK1 signaling [26] (Fig. 7.1). When rapamycin, a specific mTORC1 inhibitor, was administered to oocyte-specific PTEN mutant mice, it prevented follicular activation and preserved ovarian reserve [27]. Such an mTOR-signaling drug could offer a promising strategy for preventing POI due to overactivation of the primordial follicle pool.

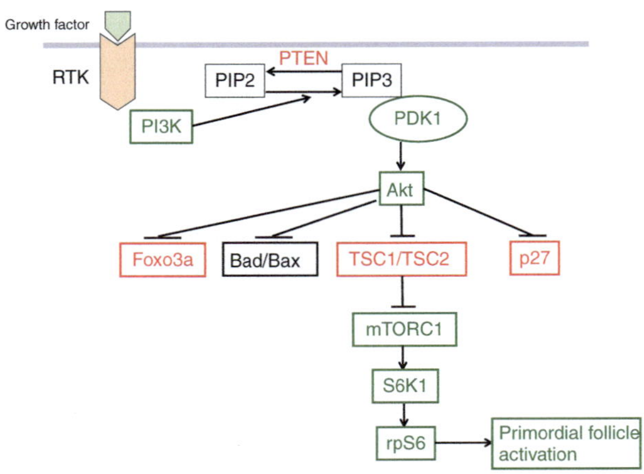

Fig. 7.1 Schema of the phosphatidylinositol 3 kinase (PI3K) pathway is shown. Molecules maintaining the dormant state are shown in red: phosphatase and tensin homolog deleted on chromosome 10 (PTEN), forkhead box O3 (Foxo3a), tuberous sclerosis complex (TSC1/TSC2), p27. Molecules maintaining the survival of the primordial follicles are shown in green: PI3K, 3-phosphoinositide-dependent kinase 1 (PDK1), Akt, mammalian target of rapamycin complex 1 (mTORC1), p70 S6 kinase 1 (S6 K1), ribosomal protein S6 (rpS6)

Somatic cells surrounding oocytes in primordial follicles also have a role in maintaining the dormant state. Morphologically, primordial follicle granulosa cells (pfGCs) exist in a flattened single-cell layer. The initiation of follicle growth starts with the proliferation of pfGCs, which then differentiate into cuboidal granulosa cells. Zhang et al. used pfGC-specific, mTORC1-signaling, model mice to show the result of inhibiting mTORC1 signaling in pfGCs [28]. They successfully showed that inhibiting mTORC1 prevented the differentiation of pfGCs into granulosa cells and that overactivation of mTORC1 signaling accelerated the differentiation of pfGCs into granulosa cells before causing premature activation of all quiescent oocytes and primordial follicles [28]. They also showed that KIT ligand expressed in pfGCs triggered the activation of dormant oocytes.

7.4 Chemotherapy-Induced Mechanism

The mechanism through which chemotherapy causes POI has been studied with in vitro and in vivo animal models. POI was to be induced by the loss of primordial follicles. However, this was not only by the direct effect of chemotherapeutic drugs on primordial follicles but also by the increased rate of growth initiation of resting primordial follicles [29].

Cyclophosphamide is an alkylating agent that is used to treat Hodgkin lymphomas and breast cancers. It results in intra- and interstrand cross-linking of DNA, which interferes with cell division. Cyclophosphamide is highly gonadotoxic and has been shown to induce ovarian damage by activating the PI3K–PTEN–Akt pathway, resulting in primordial follicle activation and follicle "burnout" [30]. Kalich-Philosoph et al. showed that co-administration of an immunomodulatory compound, AS101, reduced follicle activation and increased follicle reserve, effectively rescuing fertility after cyclophosphamide treatment [30]. Recently, mTORC1 inhibition, which is a downstream target of the PI3K–PTEN–Akt pathway, has been shown to preserve ovarian reserves, primordial follicle counts, serum anti-Müllerian hormone (AMH) levels, and fertility. This was achieved using the clinically approved drug everolimus [31].

Cisplatin is a DNA cross-linking agent used to treat sarcomas and germ cell tumors. It causes DNA damage by forming inter- and intrastrand DNA adducts. Using an in vitro culture of ovaries from 5-day-old mice, cisplatin was shown to induce an increase in p63 protein in oocytes [32]. Moreover, prolonged cisplatin treatment resulted in terminal dUTP nick-end labeling (TUNEL)-positive oocytes, indicative of cell death. The increase in p63 protein levels was dependent on c-Abl tyrosine kinase, and administration of the c-Abl kinase inhibitor imatinib abolished p63 accumulation and cell death. In another histological analysis using an in vitro culture system of newborn mouse ovaries, cisplatin showed oocyte-specific damage, but not in surrounding granulosa cells [33]. Morgan et al. [33] also showed the protective effect of imatinib against damage by cisplatin. In other research, cisplatin-induced POI was studied in a mouse model using intraperitoneal administration [34]. Cisplatin treatment decreased PTEN levels in that study, leading to increased phosphorylation of the signaling cascade. The PTEN–Akt–FOXO3 pathway was also shown to be involved in the development of POI after cisplatin treatment.

Doxorubicin is an anthracycline that is often used in the treatment of lymphomas, leukemias, and breast cancers. It intercalates with DNA and prevents its replication and transcription [35]. In comparison with cisplatin, doxorubicin has been shown to cause only granulosa cell-specific damage [33].

7.5 Diagnosis

Women with POI usually presented with secondary amenorrhea, irregular menstruation, or infertility. Although the lack of standardized diagnostic criteria for POI can cause diagnostic uncertainty, timely diagnosis and intervention are vital to avoid an estrogen deficiency state [36]. At present, the diagnostic criteria are age less than 40 years, oligo-amenorrhea for at least 4 months, elevations in two serum FSH levels obtained 4–6 weeks apart (>40 IU/L), and a low estradiol level. It is usually recommended that gonadotropin levels be estimated 3–4 months after the onset of amenorrhea or menstrual irregularity. If elevated, a repeat FSH test should be done after 4–6 weeks to confirm hypogonadism. Transvaginal ultrasound should be performed to assess the antral follicle count. The AMH level, which is produced by developing antral follicles, is currently thought to be the most reliable measure of reduced ovarian reserve. AMH has the advantage of being measurable at any time of the menstrual cycle. Levels decline before menstrual irregularity or FSH elevations occur, and it has been shown to be useful in distinguishing patients with POI from controls [37].

7.6 Clinical Management of POI

Women affected by POI must be treated as having a complex condition with physical and psychological elements that require a multidisciplinary approach. This may include gynecologists, endocrinologists, fertility specialists, oncologists, hematologists, and psychologists. Unexpected infertility is a life-altering diagnosis for many women, and they may experience severe emotional distress. Although these women report wanting guidance on how to cope with the emotional sequelae, few appear to ask for it directly [6]. Menopause is known to affect cognitive function adversely [38], with evidence that surgical menopause causes distinct impairment of cognitive function and memory [39, 40]. POI is also associated with an increased risk of osteoporosis and cardiovascular disease [41, 42].

The average serum estradiol level during the menstrual cycle is approximately 100 pg/mL in women who have regular menstruation [43]. Although no studies have compared doses of hormone replacement therapy (HRT) for women with POI, a dose of 100 μg estradiol per day, administered by transdermal patch, effectively achieves average serum estradiol levels in this range and treats symptoms [6]. Transdermal, oral, or transvaginal estradiol in doses of 100 μg daily is the therapy of choice to mimic normal physiology. Transdermal estradiol has less risk of causing thromboembolism compared with oral administration due to the first-pass effect; therefore, transdermal HRT may be preferred in women with coagulation disturbances or who are prone to thrombosis (e.g., patients with thrombophilias) [44]. The addition of cyclic progesterone for 10–12 days each month provides protection against endometrial hyperplasia and endometrial cancer [45]. Oral contraceptives are not recommended first line because they use supraphysiologic doses [6].

Women with POI have been shown to suffer reduced bone mineral density compared with healthy controls [46]. To maintain density, a healthy lifestyle is required, which will involve reductions in smoking, alcohol, and caffeine use. The North American Menopause Society recommends taking 1200 mg calcium per day and maintaining an adequate vitamin D level, defined as a serum 25-hydroxyvitamin D level of 30 ng/mL [47]. It has been recommended that adults with inadequate sunlight exposure take 800–1000 IU of vitamin D daily [48]. Regular physical activities, such as jogging, walking, and resistance exercises, are encouraged [49].

Pregnancies might occur in women with POI even when few functional follicles are present. For example, when high levels of FSH are suppressed by ethinylestradiol or gonadotropin-releasing hormone analogs, followed by ovarian induction, there is a chance of pregnancy [50]. Typical options in the face of POI include adoption, fostering, egg donation, and embryo donation. Adoption might not be acceptable to some patients, and oocyte donation may be the treatment of choice; however, religious and legal restrictions exist in some countries. It should be noted that women with POI who become pregnant after oocyte donation may be at increased risk of pregnancy-induced hypertensive disorders [51].

Fertility preservation is an important issue for women of reproductive age diagnosed with cancer who are treated with ovariotoxic chemotherapy or radiation therapy. Advances in cryopreservation techniques for oocytes and ovarian tissue have made it possible for these women to preserve fertility. Supporting this, ovarian tissue transplantation is growing rapidly, resulting in more than 70 live births worldwide [52]. Based on studies of oocyte-specific deletion of PTEN in mutant mice, short-term in vitro activation of dormant ovarian follicles after stimulation of the PTEN–Akt pathway has been demonstrated to be capable of generating a large number of mature eggs [53]. In that research, a PTEN inhibitor and PI3K stimulator were used to activate dormant primordial follicles, before transplanting them into ovariectomized recipient mice. It has been shown that disruption of the Hippo signaling pathway by ovarian tissue fragmentation in the ovary could also promote follicle growth [54]. Thus, the combination of ovarian cryopreservation, fragmentation, and in vitro activation, followed by autotransplantation, could be developed to treat infertility in women with POI [55].

The use of gonadotropin-releasing hormone analogs (GnRHa) to protect ovarian function during chemotherapy has been studied for many years, but with inconsistent results. Recent studies in patients with breast cancers and lymphomas produced controversial outcomes. The Prevention of Early Menopause Study, a large randomized controlled trial of patients with hormone-receptor-negative breast cancer [56], showed that co-administration of GnRHa reduced ovarian failure rates, increased pregnancy rates, and improved disease-free survival and overall survival rates. However, a recent trial in patients with lymphoma showed that there was no benefit in the prevention of POI [57]. Although the use of GnRHa remains controversial, it should be considered for women of reproductive age, provided there are no contraindications, before they undergo gonadotoxic chemotherapy.

Conclusion

Women who are given a diagnosis of POI suffer a wide range of physiological and psychological effects, including impaired fertility. Therefore, a multidisciplinary approach is required for the life-long management of POI. Appropriate HRT is essential both for the relief of the symptoms and for the maintenance of future bone, cardiovascular, and neuropsychological health. Ovarian function has been known to recover and result in pregnancy in women with limited POI. However, if diagnosed early or if POI can be foreseen, most women should consider freezing their oocytes or eggs to keep their options for future pregnancy open.

Acknowledgments The author would like to thank Enago (www.enago.jp) for the English language review.

Disclosures
Conflict of Interest: Yoshinori Okamura has nothing to disclose.

References

1. Albright F, Smith P, Fraser R. A syndrome characterized by primary ovarian failure. Am J Med Sci. 1942;204:625–48.
2. Welt C. Primary ovarian insufficiency: a more accurate term for premature ovarian failure. Clin Endocrinol. 2008;68:499–509.
3. Coulam CB, Adamson SC, Annegers JF. Incidence of premature ovarian failure. Obstet Gynecol. 1986;67:604–6.
4. Nelson L, Covington S, Rebar R. An update: spontaneous premature ovarian failure is not an early menopause. Fertil Steril. 2005;83:1327–32.
5. Morabia A, Costanza M, The World Health Organization Collaborative Study of Neoplasia and Steroid Contraceptives. International variability in ages at menarche, first livebirth, and menopause. Am J Epidermiol. 1998;148:1195–205.
6. Nelson LM. Primary ovarian insufficiency. N Engl J Med. 2009;360:606–14.
7. De Vos M, Devroey P, Fauser BC. Primary ovarian insufficiency. Lancet. 2010;376:911–21.
8. Sybert VP, McCauley E. Turner' s syndrome. N Engl J Med. 2004;351:1227–38.
9. Garber K, Smith K, Reines D, Warren S. Transcription, translation and fragile X syndrome. Curr Opin Genet Dev. 2006;16:270–5.
10. Janse F, Knauff EA, Niermeijer MF, Eijkemans MJ, Laven JS, Lambalk CB, Fauser BC, Goverde AJ, Dutch Premature Ovarian Failure Consortium. Similar phenotype characteristics comparing familial and sporadic premature ovarian failure. Menopause. 2010;17:758–65.
11. Jiao X, Qin C, Li J, Qin Y, Gao X, Zhang B, Zhen X, Feng Y, Simpson JL, Chen ZJ. Cytogenetic analysis of 531 Chinese women with premature ovarian failure. Hum Reprod. 2012;27:2201–7.
12. Marozzi A, Vegetti W, Manfredini E, Tibiletti MG, Testa G, Crosignani PG, Ginelli E, Meneveri R, Dalprà L. Association between idiopathic premature ovarian failure and fragile X premutation. Hum Reprod. 2000;15:197–202.

13. Willemsen R, Levenga J, Oostra BA. CGG repeat in the FMR1 gene: size matters. Clin Genet. 2011;80:214–25.
14. Wittenberger MD, Hagerman RJ, Sherman SL, McConkie-Rosell A, Welt CK, Rebar RW, Corrigan EC, Simpson JL, Nelson LM. The FMR1 premutation and reproduction. Fertil Steril. 2007;87:456–65.
15. Otsuka F, McTavish KJ, Shimasaki S. Integral role of GDF-9 and BMP-15 in ovarian function. Mol Reprod Dev. 2011;78:9–21.
16. Rafique S, Sterling E, Lawrence N. A new approach to primary ovarian insufficiency. Obstet Gynecol Clin N Am. 2012;39:567–86.
17. Castrillon DH, Miao L, Kollipara R, Horner JW, DePinho RA. Suppression of ovarian follicle activation in mice by the transcription factor Foxo3a. Science. 2003;301:215–8.
18. Lan ZJ, Xu X, Cooney AJ. Differential oocyte-specific expression of Cre recombinase activity in GDF-9-iCre, Zp3cre, and Msx2Cre transgenic mice. Biol Reprod. 2004;71:1469–74.
19. Reddy P, Liu L, Adhikari D, Jagarlamudi K, Rajareddy S, Shen Y, Du C, Tang W, Hämäläinen T, Peng SL, Lan ZJ, Cooney AJ, Huhtaniemi I, Liu K. Oocyte-specific deletion of Pten causes premature activation of the primordial follicle pool. Science. 2008;319:611–3.
20. Reddy P, Adhikari D, Zheng W, Liang S, Hämäläinen T, Tohonen V, Ogawa W, Noda T, Volarevic S, Huhtaniemi I, Liu K. PDK1 signaling in oocytes controls reproductive aging and lifespan by manipulating the survival of primordial follicles. Hum Mol Genet. 2009;18:2813–24.
21. Zoncu R, Efeyan A, Sabatini DM. mTOR: from growth signal integration to cancer, diabetes and ageing. Nat Rev Mol Cell Biol. 2011;12:21–35.
22. Thomas G. The S6 kinase signaling pathway in the control of development and growth. Biol Res. 2002;35:305–13.
23. Pullen N, Dennis PB, Andjelkovic M, Dufner A, Kozma SC, Hemmings BA, Thomas G. Phosphorylation and activation of p70s6k by PDK1. Science. 1998;279:707–10.
24. Efeyan A, Zoncu R, Sabatini DM. Amino acids and mTORC1: from lysosomes to disease. Trends Mol Med. 2012;18:524–33.
25. Adhikari D, Zheng W, Shen Y, Gorre N, Hämäläinen T, Cooney AJ, Huhtaniemi I, Lan ZJ, Liu K. Tsc/mTORC1 signaling in oocytes governs the quiescence and activation of primordial follicles. Hum Mol Genet. 2010;19:397–410.
26. Reddy P, Zheng W, Liu K. Mechanisms maintaining the dormancy and survival of mammalian primordial follicles. Trends Endocrinol Metab. 2010;21:96–103.
27. Adhikari D, Risal S, Liu K, Shen Y. Pharmacological inhibition of mTORC1 prevents over-activation of the primordial follicle pool in response to elevated PI3K signaling. PLoS One. 2013;8:e53810.
28. Zhang H, Risal S, Gorre N, Busayavalasa K, Li X, Shen Y, Bosbach B, Brännström M, Liu K. Somatic cells initiate primordial follicle activation and govern the development of dormant oocytes in mice. Curr Biol. 2014;24:2501–8.
29. Morgan S, Anderson RA, Gourley C, Wallace WH, Spears N. How do chemotherapeutic agents damage the ovary? Hum Reprod Update. 2012;18:525–35.
30. Kalich-Philosoph L, Roness H, Carmely A, Fishel-Bartal M, Ligumsky H, Paglin S, Wolf I, Kanety H, Sredni B, Meirow D. Cyclophosphamide triggers follicle activation and "burnout"; AS101 prevents follicle loss and preserves fertility. Sci Transl Med. 2013;5:185ra62.
31. Goldman KN, Chenette D, Arju R, Duncun FE, Keefe DJ, Grifo JA, Schneider RJ. mTORC1/2 inhibition preserves ovarian function and fertility during genotoxic chemotherapy. Proc Natl Acad Sci U S A. 2017;114:3186–91.
32. Gonfloni S, Di Tella L, Caldarola S, Cannata SM, Klinger FG, Di Bartolomeo C, Mattei M, Candi E, De Felici M, Melino G, Cesareni G. Inhibition of the c-Abl-TAp63 pathway protects mouse oocytes from chemotherapy-induced death. Nat Med. 2009;15:1179–85.
33. Morgan S, Lopes F, Gourley C, Anderson RA, Spears N. Cisplatin and doxorubicin induce distinct mechanisms of ovarian follicle loss; imatinib provides selective protection only against cisplatin. PLoS One. 2013;8:e70117.
34. Chang EM, Lim E, Yoon S, Jeong K, Bae S, Lee DR, Yoon TK, Choi Y, Lee WS. Cisplatin Induces overactivation of the dormant primordial follicle through PTEN/AKT/FOXO3a pathway which leads to loss of ovarian reserve in mice. PLoS One. 2015;10:e0144245.

35. Jurisicova A, Lee HJ, D'Estaing SG, Tilly J, Perez GI. Molecular requirements for doxorubicin-mediated death in murine oocytes. Cell Death Differ. 2006;13:1466–74.
36. Maclaran K, Panay N. Current concepts in premature ovarian insufficiency. Womens Health. 2015;11:169–82.
37. Sahmay S, Usta TA, Erel T, Atakul N, Aydogan B. Elevated LH levels draw a stronger distinction than AMH in premature ovarian insufficiency. Climacteric. 2014;17:197–203.
38. Kok HS, Kuh D, Cooper R, van der Schouw YT, Grobbee DE, Wadsworth ME, Richards M. Cognitive function across the life course and the menopausal transition in a British birth cohort. Menopause. 2006;13:19–27.
39. Farrag AK, Khedr EM, Abdel-Aleem H, Rageh TA. Effect of surgical menopause on cognitive functions. Dement Geriatr Cogn Disord. 2002;13:193–8.
40. Nappi RE, Sinforiani E, Mauri M, Bono G, Polatti F, Nappi G. Memory functioning at menopause: impact of age in ovariectomized women. Gynecol Obstet Investig. 1999;47:29–36.
41. Gallagher JC. Effect of early menopause on bone mineral density and fractures. Menopause. 2007;14:567–71.
42. Jacobsen BK, Knutsen SF, Fraser GE. Age at natural menopause and total mortality and mortality from ischemic heart disease: the Adventist Health Study. J Clin Epidemiol. 1999;52:303–7.
43. Mishell DR Jr, Nakamura RM, Crosignani PG, Stone S, Kharma K, Nagata Y, Thorneycroft IH. Serum gonadotropin and steroid patterns during the normal menstrual cycle. Am J Obstet Gynecol. 1971;111:60–5.
44. Scarabin PY, Oger E, Plu-Bureau G, EStrogen and THromboEmbolism Risk Study Group. Differential association of oral and transdermal oestrogen-replacement therapy with venous thromboembolism risk. Lancet. 2003;362:428–32.
45. Primary ovarian insufficiency in adolescent and young women. Committee opinion. No 605. ACOG Obstet Gynecol. 2014;123:193–7.
46. Anasti JN, Kalantaridou SN, Kimzey LM, Defensor RA, Nelson LM. Bone loss in young women with karyotypically normal spontaneous premature ovarian failure. Obstet Gynecol. 1998;91:12–5.
47. North American Menopause Society. The role of calcium in peri- and postmenopausal women: 2006 position statement of the North American Menopause Society. Menopause. 2006;13:862–77.
48. Holick MF. Vitamin D deficiency. N Engl J Med. 2007;357:266–81.
49. Martyn-St James M, Carroll S. Meta-analysis of walking for preservation of bone mineral density in postmenopausal women. Bone. 2008;43:521–31.
50. Check JH. Mild ovarian stimulation. J Assist Reprod Genet. 2007;24:621–7.
51. Salha O, Sharma V, Dada T, Nugent D, Rutherford AJ, Tomlinson AJ, Philips S, Allgar V, Walker JJ. The influence of donated gametes on the incidence of hypertensive disorders of pregnancy. Hum Reprod. 1999;14:2268–73.
52. Silber S. Ovarian tissue cryopreservation and transplantation: scientific implications. J Assist Reprod Genet. 2016;33:1595–603.
53. Li J, Kawamura K, Cheng Y, Liu S, Klein C, Liu S, Duan EK, Hsueh AJ. Activation of dormant ovarian follicles to generate mature eggs. Proc Natl Acad Sci U S A. 2010;107:10280–4.
54. Kawamura K, Cheng Y, Suzuki N, Deguchi M, Sato Y, Takae S, Ho CH, Kawamura N, Tamura M, Hashimoto S, Sugishita Y, Morimoto Y, Hosoi Y, Yoshioka N, Ishizuka B, Hsueh AJ. Hippo signaling disruption and Akt stimulation of ovarian follicles for infertility treatment. Proc Natl Acad Sci U S A. 2013;110:17474–9.
55. Suzuki N, Yoshioka N, Takae S, Sugishita Y, Tamura M, Hashimoto S, Morimoto Y, Kawamura K. Successful fertility preservation following ovarian tissue vitrification in patients with primary ovarian insufficiency. Hum Reprod. 2015;30:608–15.
56. Moore HC, Unger JM, Phillips KA, Boyle F, Hitre E, Porter D, Francis PA, Goldstein LJ, Gomez HL, Vallejos CS, Partridge AH, Dakhil SR, Garcia AA, Gralow J, Lombard JM, Forbes JF, Martino S, Barlow WE, Fabian CJ, Minasian L, Meyskens FL Jr, Gelber RD, Hortobagyi GN, Albain KS, POEMS/S0230. Goserelin for ovarian protection during breast-cancer adjuvant chemotherapy. N Engl J Med. 2015;372:923–32.

57. Demeestere I, Brice P, Peccatori FA, Kentos A, Dupuis J, Zachee P, Casasnovas O, Van Den Neste E, Dechene J, De Maertelaer V, Bron D, Englert Y. No evidence for the benefit of gonadotropin-releasing hormone agonist in preserving ovarian function and fertility in lymphoma survivors treated with chemotherapy: final long-term report of a prospective randomized trial. J Clin Oncol. 2016;34:2568–74.

Rescue for Primary Ovarian Insufficiency 8

Kyu Sup Lee

Abstract

Advancing female age causes the changes in ovarian function and finally results in reproductive aging. Ovarian aging is characterized by a gradual reduction of follicle number and oocyte quality. Oocyte aging also occurs in the ovaries of reproductive aged female, and ovarian aging is directly associated with a poor fertilizing/developmental competency of oocytes. Therefore treatments to reverse oocyte aging have been a significant issue in assisted reproduction technologies (ART). Recently, inadequate ovarian angiogenesis has been suggested to result in age-related damage of oocytes. A potential therapeutic way which improves poor oocyte quality due to ovarian aging is suggested by the activation of ovarian angiogenesis.

Keywords

Ovarian aging · Oocyte aging · Oocyte quality · Angiogenesis

8.1 Introduction

Nowadays first childbearing has been postponed to advanced ages by many women due to the availability of contraceptive methods, a growing economical wealth through the opportunity to increase their level of education and to participate in the labor force [1, 2]. This social tendency to postpone childbearing has made reproductive aging and it has been a main cause for increasing infertile couple by age factor [3]. In general, female fertility starts to decline significantly in the early 30s and to

K. S. Lee
Department of Obstetrics and Gynecology, Medical Research Institute, Pusan National University School of Medicine, Busan, South Korea
e-mail: kuslee@pusan.ac.kr

H. Katabuchi et al. (eds.), *Cell Biology of the Ovary*,
https://doi.org/10.1007/978-981-10-7941-2_8

decrease steeply after age 35, culminating in the menopause at 50–51 years of age [4]. However, advancing female age remains a fastidious problem in infertility treatment.

The ovary is the main organ of female fertility, and it is crucially affected by natural aging more than any other tissues. The process of the age-related physiological reproductive aging in female is very closely associated with the changes of ovarian function. Therefore, the aged ovary has characterized by a gradual reduction of the primordial follicles in the ovarian cortex with a biphasic pattern, especially before and after the age of 38 [5–7]. Menopause is the final step in the process of ovarian aging.

The reproductive aged ovary results to oocyte aging [8]. The matured oocytes are normally fertilized within an optimal fertilization time after ovulation. However, if the oocytes are not fertilized within that time, the quality of unfertilized oocytes will be deteriorated in a time-dependent manner. This process is called "oocyte aging," which gradually decreases the quality and quantity of oocytes, including decreased number of ovulated oocytes, decreased viability of preimplantation embryos, and increased percentages of abnormal/degenerating oocytes [9, 10]. In this respect, ovarian aging is directly associated with a poor fertilizing/developmental competency of aged oocytes, and it becomes a common cause of assisted reproduction technology (ART) failures [11].

Most studies on reproductive aging have focused on the quantitative aspect of follicle loss as a key determinant of ovarian aging. They did not take into account the inactivation or inadequacy of other biological processes [12]. However, oocyte quality and quantity with the age seems to be affected by several factors including endocrine, paracrine, genetic, and metabolic factors, although it is still not well defined. Ovarian angiogenesis plays an important role in a series of event of folliculogenesis, and active blood supply may be essential for the production of good-quality oocyte [13]. Therefore, nowadays, in order to understand the mechanisms of ovarian aging and oocyte aging, the research on genomic changes and the whole ovarian microenvironment, such as ovarian angiogenesis, has gained importance. This review focuses on the role of ovarian angiogenesis in ovarian and oocyte aging, and we also discuss a potential therapeutic way which recovers poor oocyte quality due to ovarian aging by the activation of ovarian angiogenesis.

8.2 Assessment of Ovarian Aging

1. Assessment of antral follicle count by ultrasonography

   Ovarian aging has been supposed to be closely related to an ongoing decline in the number of primordial follicles. The number of primordial follicles in the ovary has a correlation with the number of growing follicle, and it certainly leads to a decrease in the number of the antral follicles [14]. Ovarian antral follicles can be easily detected by transvaginal ultrasonography, which has an advantage of noninvasive assessment. Therefore, several studies have demonstrated that antral follicle count has the clearest correlation with age, and it has been used as a predictive marker for ovarian reserve in the patients undergoing in vitro fertilization treatment [15, 16].

2. Hormonal assessment of anti-müllerian hormone (AMH)

   The development and growth of the antral follicles are controlled by several regulators in endocrine, paracrine, and autocrine manners. Among them, AMH has recently started to emerge as an important regulator of mammalian follicular development [17, 18] because it is expressed at low levels in primary follicles, increases to maximal levels in large preantral and small antral follicles, and then declines as the follicle grows [19, 20]. It is a dimeric glycoprotein and a member of the transforming growth factor-beta superfamily, and it is produced exclusively by granulosa cells of ovarian follicles [21, 22]. Serum AMH has been known to decline with age and is highly associated with ovarian follicular reserve and disordered folliculogenesis [23–25]. As a result, serum AMH level has been clinically used as a good biomarker to predict ovarian response in ART [20, 26, 27] and to evaluate the non-growing follicle pool and early ovarian aging [28–30].

3. Genomic assessment using cumulus cells: ovarian aging-related genes

   Studies on menopausal age of mothers and daughters and also between sisters have implicated that the genetic factors are involved in the reproductive aging process [31]. Several kinds of genes have been elucidated to play a role in ovarian aging. First are genes that exert known hormonal effects. They primarily affect follicle function and include follicle-stimulating hormone (FSH), follicle-stimulating hormone receptor (FSHR), luteinizing hormone (LH), luteinizing hormone receptor (LHR), cytochrome P450 17 (CYP17), and cytochrome P450 19 (CYP19) [32, 33]. The second genes are growth differentiation factor 9 (GDF9), bone morphogenetic protein 15 (BMP15), and G-protein-coupled receptor 3 (GPR3). These genes are known to affect the rate of initial selection from the primordial follicle pool [34, 35]. Third genes include deoxyribonucleic acid (DNA)-binding protein and transcription factors, like newborn ovary homeobox (NOBOX) and LIM homeobox protein (LHX8), and RNA-binding proteins, like NANOS [36]. Recently, the interest in mitochondrial energy production-related genes has been increased [37, 38] because oocytes from aged women may have mitochondrial dysfunction due to increased oxidative stress [39, 40]. Small variation or mutation in these genes could result in the decrease of the follicle pool [41, 42]. Like this, genomic research has an important role in the understanding of the mechanisms of ovarian aging [43]. Transcriptome profiles revealed a substantial difference between younger and older human oocytes [44].

However, it is not easy to obtain the oocytes or the ovaries in human study due to an ethical issue. In this respect, gene expression in human cumulus cells has been wildly used as a powerful noninvasive approach to evaluate not only oocyte quality and developmental competency but also ovarian aging [45–49]. This rationale is based on the characteristics of cumulus cells, which enclose the oocytes, forming the cumulus-oocyte complex (COC) and undergoing a continuous cross talk with oocyte during follicular development [50]. The differentiation and expansion of cumulus cells are controlled by the oocyte, and it regulates the metabolism of the glucose and pyruvate used for energy production in the oocyte [51].

4. Developmental competency of oocyte using in vitro culture

   Generally, ovarian aging results in decreased developmental competency of oocyte along with its aging. Therefore, if the oocytes are obtained from animal study or in vitro fertilization (IVF) practice in case of human, in vitro culture to preimplantation embryo stages using those oocytes can be the most direct approach in the assessment of oocyte aging.

8.3 Effect of Ovarian Angiogenesis on Oocyte Aging

In general, oocytes have a "temporal window for optimal fertilization," and those beyond temporal window are called to be postovulatory or postmaturation aged. These oocytes are called "aged oocytes." They are not fertilized within that temporal window time and finally lead to parthenogenetic activation and fragmentation [52, 53]. Aging-related ovarian aging also eventually leads to oocyte aging. Tatone et al. suggested that subtle oxidative damage in primordial follicles as well as in ovarian stroma vessels may be possible mechanisms underlying the process of follicle aging, considering the effect of aging on ovarian and follicular vascularization as the etiology of oxidative stress. This finding strongly implies that dysfunction in ovarian angiogenesis may be an important cause of follicles and, in turn, oocyte aging with advancing age [54].

Angiogenesis is the formation of new blood vessels from pre-existing vasculature. Generally, angiogenesis does not occur except during wound healing which requires neovascularization. However, the female reproductive tract, including the ovaries and uterus, has cyclic angiogenic processes to facilitate follicular development and implantation processes. Especially, in ovarian angiogenesis, vascularity is necessary for each step of follicular development from assembly and activation of primordial follicle, the transition to primary follicle. Vasculature of the follicle plays an important role in the preferential delivery of gonadotropin, oxygen, and nutrients essential for follicular development [13]. During the menstrual cycle, angiogenesis and vascular regression occur cyclically, and vascular density markedly increased in follicles undergoing development from preantral to antral stage [55, 56]. Gaulden proposed that a deficient microvasculature develops around the dominant follicles with aging, resulting in the hypoxia of follicles [57]. Follicles undergoing atresia have decreased vascularity during follicular development, while the selected follicles show a more elaborate microvasculature, and the mature follicles have increased vascular density [58–60]. These results mean that ovarian angiogenesis plays a critical role in various follicular development processes that require neovascularization, including follicular growth, selection of dominant follicles, inhibition of follicular atresia, ovulation, and corpus luteum formation [61, 62]. Indeed, dysfunction in ovarian angiogenesis may be an important cause of anovulation, polycystic ovarian syndrome (PCOS), pregnancy loss, and possibly infertility [61]. Therefore, it can be postulated that the activation of ovarian angiogenesis may be a good strategy for the improvement of age-related decline in oocyte quality.

8.4 Regulation of Ovarian Angiogenesis

Ovarian angiogenesis is a complex process regulated by coordinated interaction of numerous autocrine, paracrine, and endocrine factors. Vascular endothelial growth factor (VEGF) is a key regulator of angiogenesis through interaction with its receptor VEGFR2. It stimulates endothelial proliferation, migration, and survival, and it is an important survival factor for extravascular ovarian cells [63].

Blood flow indices (pulsatility index (PI), resistive index (RI)) in early follicular phase correlated negatively with the number of follicles recruited and the number of oocytes retrieved [64, 65]. Friedman et al. reported that the increase VEGF follicular fluid concentrations in women of advanced reproductive age are due to relative hypoxia in ovarian follicles [66]. Pellicer et al. observed reduced blood flow during natural cycles around the dominant follicle in poor responders compared with controls, which suggested that women with a response to exogenous gonadotropins typical for women of advanced reproductive age have a deficient follicular microcirculation even if the issue of age is not directly addressed [67]. It seems possible that an active blood supply through the increase of VEGF expression prevents the follicles from atresia [68]. Fraser and Duncan suggest that the supply of an appropriate blood vessel and the maintenance of vascular permeability in ovary are necessary for making adequate effect of gonadotropins and paracrine factors to sustain follicular growth and ovulation [13].

VEGF is identified in ovarian granulosa cells [69]. It is expressed not only in theca cells of follicle but also in the stroma of the ovary, and it also influences preantral follicle development toward the antral stage [70, 71]. Ovarian transcriptome study shows the alteration in follicular development-related gene expression. From primordial to primary follicle, the expressions of 148 genes are upregulated and VEGF is also significantly upregulated during the primordial to primary follicle transition [72]. VEGF is also upregulated in dominant follicle/selection process leading to ovulation [69, 73]. Conversely, follicle atresia is induced both by VEGF and follicular blood vessel network reduction [74–76]. These results mean that VEGF and ovarian angiogenesis may play a crucial role in early follicular development as well as late follicular development [77].

Indeed, several studies have demonstrated that enhancing VEGF expression during the follicular phase may be useful in increasing the number of predominant follicles destined for ovulation and angiogenesis [71, 78]. Direct injection of VEGF gene fragments or VEGF into the ovary increases angiogenesis and the number of follicles to be ovulated [79–81]. In contrast, suppression of follicular angiogenesis by the inhibition of VEGF is associated with the inhibition of antral follicular development and results in the prevention of ovulation [82]. Therefore, from these finding, we postulate that the induction of increase in ovarian VEGF gene expression by a potential factor or mechanism can be a possible strategy for the rescue of aged-related decline of oocyte quality.

8.5 Rescue of Oocyte Quality by the Activation of Ovarian Angiogenesis

Nitric oxide (NO) is known to be a potent vasodilator and angiogenic factor and to mediate the vascular permeability of VEGF [83]. NO also plays an important role in ovarian angiogenesis during folliculogenesis and ovulation, including in the determination of oocyte quality and embryo developmental competence [84]. Therefore, we firstly attempted to investigate whether co-injection of sodium nitroprusside (SNP) with gonadotropin during the superovulation process improved ovarian response and oocyte developmental competence in aged female mice [85]. SNP is a representative nitric oxide (NO) donor. Our study showed that SNP treatment improves ovarian response and oocyte developmental competence in aged female (Table 8.1) with increased expression of ovarian VEGF (Fig. 8.1), suggesting a possibility that the treatment of factor to activate ovarian angiogenesis rescue the oocyte quality in reproductive aged female.

So, we sequentially investigated whether the treatment of other angiogenic factors also improves the oocyte quality in reproductive aged mice using estrogen and

Table 8.1 Effects of angiogenic factors treatment on embryo development in aged mice

Angiogenic factors	Age of mice (weeks)	Concentration	No. of zygotes flushed/mouse	% of blastocyst
Sodium nitroprusside (SNP)		0 (control)	15.2	5.5
	25–27	1 μM	21.7	36.7*
		10 μM	20.2	46.8*
Visfatin		0 (control)	11.0	7.5
	26–31	10 ng/ml	10.5	33.3*
		100 ng/ml	13.3	39.5*
		500 ng/ml	10.7	46.4*

*$P < 0.05$ (vs. control). Reproduced from Lee et al. (2008) for SNP and Choi et al. (2012) for visfatin with permission [85, 86]

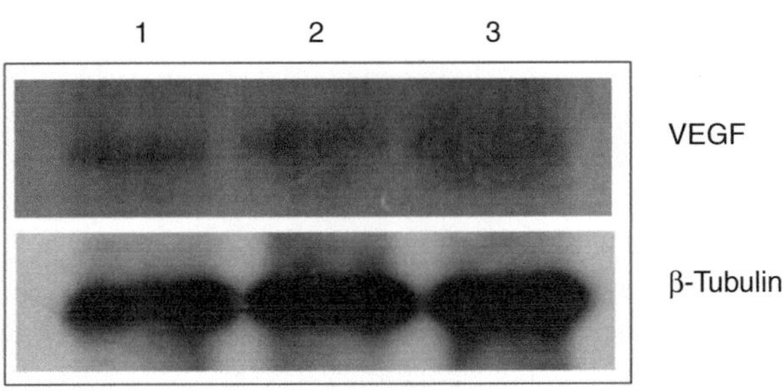

Fig. 8.1 Western blot analysis for the VEGF expression in the ovaries of aged mice (26-week old). The ovaries were isolated just after the flushing of one-cell embryos and VEGF expression was examined. Lane 1, PMSG alone (control); lane 2, 1 μM SNP + PMSG; and lane 3, 10 μM SNP + PMSG. *$P < 0.05$, 1 μM SNP and 10 μM SNP vs. control. Reproduced from Lee et al. (2008) with permission [85]

leptin [87, 88]. We have obtained the similar results although the effect has a little difference. However, these substances have some risks causing a serious disease such as cancer or heart attack.

Visfatin has recently become known as one of the adipokines released by a variety of cells, including adipose tissue [89]. It strongly correlates to visceral obesity and exerts insulin mimetic and antagonistic properties, thus resulting in a glucose-lowering effect [89–91]. Several recent studies showed that visfatin stimulates angiogenesis [92–95] and is also expressed in human granulosa cells [96]. On the basis of the characteristics of visfatin, we recently studied the effect of visfatin on the improvement of oocyte quality in reproductive aged female mice. Visfatin administration during superovulation significantly increased embryo developmental rate (Table 8.1), with the increased expression in ovarian visfatin and VEGF (Fig. 8.2). Visfatin-administered aged mice delivered significantly higher numbers

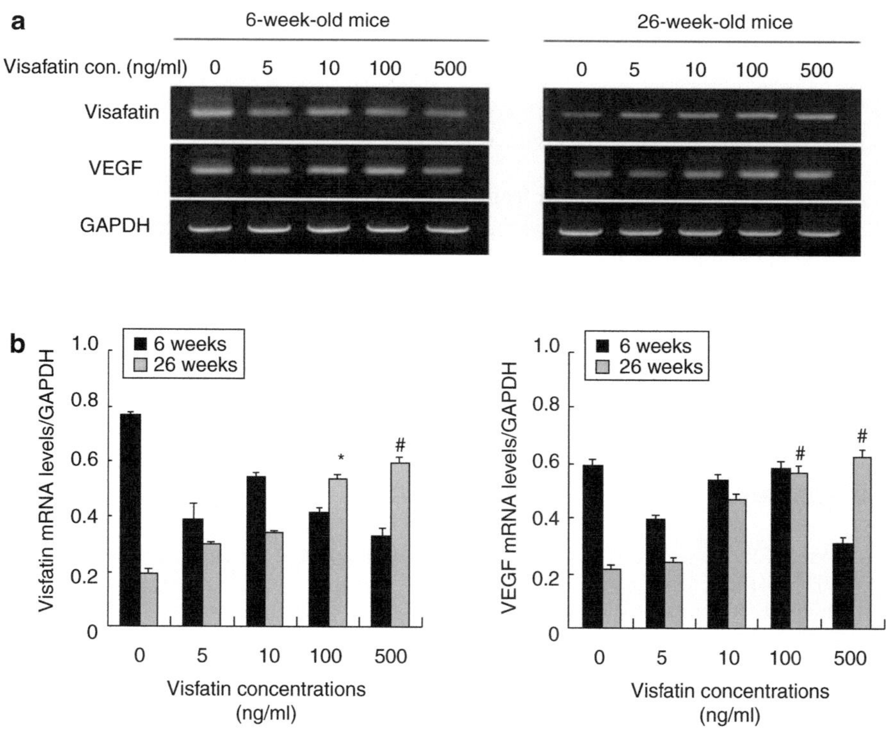

Fig. 8.2 Effect of coadministration of visfatin on expression of ovarian visfatin and VEGF in different aged mice. Various doses (0–500 ng/ml) of recombinant mouse visfatin were administered intraperitoneally to two aged mice groups (6–11-week-old and 26–31-week-old) during superovulation with PMSG and followed by hCG approximately 48 h later. Then, the mice were immediately paired with an individual male. Eighteen hours after hCG injection, both ovaries of each mouse were collected and the expressions of ovarian visfatin and VEGF were examined by RT-PCR. (**a**) Representative RT-PCR for ovarian visfatin expression of 6- and 26-week-old mice was shown. (**b**) Visfatin mRNA intensity values were normalized to those of GAPDH and expressed as means±SD; $n = 3$; 6 ovaries/age group; $^{\#}P < 0.05$; $^{*}P < 0.001$ vs. control (0 ng/ml). Reproduced from Choi et al. (2012) with permission [86]

Table 8.2 Effect of visfatin on fertility outcome in aged mice

Age of female mice	Visfatin concentration	No. of offspring/female mice mated
29–31 weeks	0	0[a]
	10 ng/ml	5.50±0.57[b]
	100 ng/ml	8.00±1.41[c]
	500 ng/ml	8.00±0.81[c]

[a]P<.05 (vs. 0 ng/ml)
[b]P<.05 (vs. 10 ng/ml)
[c]P<.01 (vs 0 ng/ml)
$P < 0.05$ (b:a, c:b), $P < 0.01$ (c:a). Mean ± SD. Reproduced from Choi et al. (2012) with permission [85, 86]

of offspring than controls (Table 8.2) [86]. This study suggests that visfatin administration during superovulation plays an important role in regulating oocyte quality and can improve oocyte quality and fertility of aged female mice.

Above all these results strongly indicate that active blood flow may be a necessary factor for an adequate folliculogenesis and the activation of ovarian angiogenesis may be a good strategy for the improvement of age-related decline in oocyte quality.

Conclusions

Oocyte aging can be resulted from mainly female aging-related ovarian aging. An inadequate activation of ovarian angiogenesis seems to be closely associated with oocyte quality decay due to this oocyte aging. Therefore this review shows the activation of ovarian angiogenesis can rescue female reproductive age-related oocyte aging, suggesting that ovarian angiogenesis may be a good target for a potential clinical implication in the treatment of age-related decline in fertility for human beings.

References

1. te Velde ER, Pearson PL. The variability of female reproductive ageing. Hum Reprod Update. 2002;8:141–54.
2. Leridon H. Demographic effects of the introduction of steroid contraception in developed countries. Hum Reprod Update. 2006;12:603–16.
3. Balasch J, Gratacós E. Delayed childbearing: effects on fertility and the outcome of pregnancy. Fetal Diagn Ther. 2011;29:263–73.
4. Broekmans FJ, Knauff EA, te Velde ER, Macklon NS, Fauser BC. Female reproductive ageing: current knowledge and future trends. Trends Endocrinol Metab. 2007;18:58–65.
5. Faddy M. Follicle dynamics during ovarian ageing. Mol Cell Endocrinol. 2000;163:43–8.
6. De Bruin J, Dorland M, Spek E, Posthuma G, Van Haaften M, Looman C, et al. Age-related changes in the ultrastructure of the resting follicle pool in human ovaries. Biol Reprod. 2004;70:419–24.
7. Broekmans F, Soules M, Fauser B. Ovarian aging: mechanisms and clinical consequences. Endocr Rev. 2009;30:465–93.
8. Tatone C, Amicarelli F. The aging ovary—the poor granulosa cells. Fertil Steril. 2013;99:12–7.
9. Tarín JJ, Pérez-Albalá S, Cano A. Cellular and morphological traits of oocytes retrieved from aging mice after exogenous ovarian stimulation 1. Biol Reprod. 2001;65:141–50.

10. Thouas GA, Trounson AO, Jones GM. Effect of female age on mouse oocyte developmental competence following mitochondrial injury. Biol Reprod. 2005;73:366–73.
11. Wilcox AJ, Weinberg CR, Baird DD. Post-ovulatory ageing of the human oocyte and embryo failure. Hum Reprod. 1998;13:394–7.
12. Hansen KR, Knowlton NS, Thyer AC, Charleston JS, Soules MR, Klein NA. A new model of reproductive aging: the decline in ovarian non-growing follicle number from birth to menopause. Hum Reprod. 2008;23:699–708.
13. Fraser HM. Regulation of the ovarian follicular vasculature. Reprod Biol Endocrinol. 2006;4:18.
14. Gougeon A, Ecochard R, Thalabard JC. Age-related changes of the population of human ovarian follicles: increase in the disappearance rate of non-growing and early-growing follicles in aging women. Obstet Gynecol Surv. 1995;50:50–3.
15. Scheffer GJ, Broekmans FJ, Dorland M, Habbema JD, Looman CW, te Velde ER. Antral follicle counts by transvaginal ultrasonography are related to age in women with proven natural fertility. Fertil Steril. 1999;72:845–51.
16. Christianson MS, Shoham G, Tobler KJ, Zhao Y, Cordeiro CN, Leong M, et al. Measurement of antral follicle count in patients undergoing in vitro fertilization treatment: results of a worldwide web-based survey. J Assist Reprod Genet. 2015;32:1435–40.
17. Knight PG, Glister C. Local roles of TGF-β superfamily members in the control of ovarian follicle development. Anim Reprod Sci. 2003;78:165–83.
18. Seifer DB, MacLaughlin DT. Mullerian inhibiting substance is an ovarian growth factor of emerging clinical significance. Fertil Steril. 2007;88:539–46.
19. Knauff EA, Eijkemans MJ, Lambalk CB, ten Kate-Booij MJ, Hoek A, Beerendonk CC, et al. Anti-Mullerian hormone, inhibin B, and antral follicle count in young women with ovarian failure. J Clin Endocrinol Metabol. 2009;94:786–92.
20. Kallio S, Aittomäki K, Piltonen T, Veijola R, Liakka A, Vaskivuo T, et al. Anti-Müllerian hormone as a predictor of follicular reserve in ovarian insufficiency: special emphasis on FSH-resistant ovaries. Hum Reprod. 2012;27:854–60.
21. Cate R, Mattaliano R, Hession C, Tizard R, Farber N, Cheung A, et al. Isolation of the bovine and human genes for Müllerian inhibiting substance and expression of the human gene in animal cells. Cell. 1986;45:685–98.
22. Kokcu A. Premature ovarian failure from current perspective. Gynecol Endocrinol. 2010;26:555–62.
23. Nelson SM, Anderson RA, Broekmans FJ, Raine-Fenning N, Fleming R, La Marca A. Anti-Müllerian hormone: clairvoyance or crystal clear? Hum Reprod. 2012;27:631–6.
24. Kedem A, Haas J, Geva LL, Yerushalmi G, Gilboa Y, Kanety H, et al. Ongoing pregnancy rates in women with low and extremely low AMH levels. A multivariate analysis of 769 cycles. PLoS One. 2013;8:e81629.
25. Jeppesen J, Anderson R, Kelsey T, Christiansen SL, Kristensen S, Jayaprakasan K, et al. Which follicles make the most anti-Müllerian hormone in humans? Evidence for an abrupt decline in AMH production at the time of follicle selection. Mol Hum Reprod. 2013;19:519–27.
26. Tehrani FR, Solaymani-Dodaran M, Tohidi M, Gohari MR, Azizi F. Modeling age at menopause using serum concentration of anti-mullerian hormone. J Clin Endocrinol Metabol. 2013;98:729–35.
27. Lin P-Y, Huang F-J, Kung F-T, Chiang H-J, Lin Y-J, Lin Y-C, et al. Evaluation of serum anti-mullerian hormone as a biomarker of early ovarian aging in young women undergoing IVF/ICSI cycle. Int J Clin Exp Pathol. 2014;7:6245.
28. Freeman EW, Sammel MD, Lin H, Gracia CR. Anti-mullerian hormone as a predictor of time to menopause in late reproductive age women. J Clin Endocrinol Metabol. 2012;97:1673–80.
29. Broer S, Eijkemans M, Scheffer G, Van Rooij I, De Vet A, Themmen A, et al. Anti-Müllerian hormone predicts menopause: a long-term follow-up study in normoovulatory women. J Clin Endocrinol Metabol. 2011;96:2532–9.
30. Dolleman M, Depmann M, Eijkemans M, Heimensem J, Broer S, van der Stroom E, et al. Anti-Müllerian hormone is a more accurate predictor of individual time to menopause than mother's age at menopause. Hum Reprod. 2014;29:584–91.

31. Wicks J, Treloar SA, Martin NG. Using identity-by-descent information in affected sib pairs to increase the efficiency of genetic association studies. Twin Res. 2004;7:211–6.
32. Zerbetto I, Gromoll J, Luisi S, Reis FM, Nieschlag E, Simoni M, et al. Follicle-stimulating hormone receptor and DAZL gene polymorphisms do not affect the age of menopause. Fertil Steril. 2008;90:2264–8.
33. Small CM, Marcus M, Sherman SL, Sullivan AK, Manatunga AK, Feigelson HS. CYP17 genotype predicts serum hormone levels among pre-menopausal women. Hum Reprod. 2005;20:2162–7.
34. Andreu-Vieyra C, Lin Y-N, Matzuk MM. Mining the oocyte transcriptome. Trends Endocrinol Metab. 2006;17:136–43.
35. Gilchrist RB, Lane M, Thompson JG. Oocyte-secreted factors: regulators of cumulus cell function and oocyte quality. Hum Reprod Update. 2008;14:159–77.
36. Skillern A, Rajkovic A. Recent developments in identifying genetic determinants of premature ovarian failure. Sex Dev. 2008;2:228–43.
37. Perez GI, Jurisicova A, Matikainen T, Moriyama T, Kim M-R, Takai Y, et al. A central role for ceramide in the age-related acceleration of apoptosis in the female germline. FASEB J. 2005;19:860–2.
38. Tatone C, Carbone M, Falone S, Aimola P, Giardinelli A, Caserta D, et al. Age-dependent changes in the expression of superoxide dismutases and catalase are associated with ultrastructural modifications in human granulosa cells. Mol Hum Reprod. 2006;12:655–60.
39. Agarwal A, Gupta S, Sharma RK. Role of oxidative stress in female reproduction. Reprod Biol Endocrinol. 2005;3:28.
40. Eichenlaub-Ritter U, Wieczorek M, Lüke S, Seidel T. Age related changes in mitochondrial function and new approaches to study redox regulation in mammalian oocytes in response to age or maturation conditions. Mitochondrion. 2011;11:783–96.
41. Kok HS, van Asselt KM, van der Schouw YT, Peeters PH, Wijmenga C. Genetic studies to identify genes underlying menopausal age. Hum Reprod Update. 2005;11:483–93.
42. Wang T-T, Ke Z-H, Song Y, Chen L-T, Chen X-J, Feng C, et al. Identification of a mutation in GDF9 as a novel cause of diminished ovarian reserve in young women. Hum Reprod. 2013;28:2473–81.
43. Şükür YE, Kıvançlı İB, Özmen B. Ovarian aging and premature ovarian failure. J Turk Ger Gynecol Assoc. 2014;15:190.
44. Grøndahl M, Andersen CY, Bogstad J, Nielsen F, Meinertz H, Borup R. Gene expression profiles of single human mature oocytes in relation to age. Hum Reprod. 2010;25:957–68.
45. Assou S, Haouzi D, Mahmoud K, Aouacheria A, Guillemin Y, Pantesco V, et al. A non-invasive test for assessing embryo potential by gene expression profiles of human cumulus cells: a proof of concept study. Mol Hum Reprod. 2008;14:711–9.
46. Assou S, Haouzi D, De Vos J, Hamamah S. Human cumulus cells as biomarkers for embryo and pregnancy outcomes. Mol Hum Reprod. 2010;16:531–8.
47. Assou S, Boumela I, Haouzi D, Anahory T, Dechaud H, De Vos J, et al. Dynamic changes in gene expression during human early embryo development: from fundamental aspects to clinical applications. Hum Reprod Update. 2010;17:272–90.
48. McReynolds S, Dzieciatkowska M, McCallie BR, Mitchell SD, Stevens J, Hansen K, et al. Impact of maternal aging on the molecular signature of human cumulus cells. Fertil Steril. 2012;98:1574–80. e5.
49. Pacella L, Zander-Fox DL, Armstrong DT, Lane M. Women with reduced ovarian reserve or advanced maternal age have an altered follicular environment. Fertil Steril. 2012;98:986–94. e2.
50. Albertini DF, Combelles C, Benecchi E, Carabatsos MJ. Cellular basis for paracrine regulation of ovarian follicle development. Reproduction. 2001;121:647–53.
51. Russell DL, Robker RL. Molecular mechanisms of ovulation: co-ordination through the cumulus complex. Hum Reprod Update. 2007;13:289–312.
52. Goud A, Goud P, Diamond M, Abu-soud H. Nitric oxide delays oocyte aging. Free Radic Biol Med. 2004;37:S81.

53. Miao Y-L, Kikuchi K, Sun Q-Y, Schatten H. Oocyte aging: cellular and molecular changes, developmental potential and reversal possibility. Hum Reprod Update. 2009;15:573–85.
54. Tatone C, Amicarelli F, Carbone MC, Monteleone P, Caserta D, Marci R, et al. Cellular and molecular aspects of ovarian follicle ageing. Hum Reprod Update. 2008;14:131–42.
55. Reynolds LP, Killilea S, Redmer D. Angiogenesis in the female reproductive system. FASEB J. 1992;6:886–92.
56. Suzuki T, Sasano H, Takaya R, Fukaya T, Yajima A, Nagura H. Cyclic changes of vasculature and vascular phenotypes in normal human ovaries. Hum Reprod. 1998;13:953–9.
57. Gaulden ME. Maternal age effect: the enigma of Down syndrome and other trisomic conditions. Mutat Res. 1992;296:69–88.
58. Ravindranath N, Little-Ihrig L, Phillips HS, Ferrara N, Zeleznik AJ. Vascular endothelial growth factor messenger ribonucleic acid expression in the primate ovary. Endocrinology. 1992;131:254–60.
59. Bedaiwy MA, Hussein MR, Biscotti C, Falcone T. Cryopreservation of intact human ovary with its vascular pedicle. Hum Reprod. 2006;21:3258–69.
60. Martinez-Madrid B, Donnez J. Cryopreservation of intact human ovary with its vascular pedicle—or cryopreservation of hemiovaries? Hum Reprod. 2007;22:1795–6.
61. Geva E, Jaffe RB. Role of vascular endothelial growth factor in ovarian physiology and pathology. Fertil Steril. 2000;74:429–38.
62. Redmer DA, Reynolds LP. Angiogenesis in the ovary. Rev Reprod. 1996;1:182–92.
63. Greenaway J, Connor K, Pedersen HG, Coomber BL, LaMarre J, Petrik J. Vascular endothelial growth factor and its receptor, Flk-1/KDR, are cytoprotective in the extravascular compartment of the ovarian follicle. Endocrinology. 2004;145:2896–905.
64. Altundag M, Levi R, Adakan S, Goker EN, Killi R, Ozcakir HT, et al. Intraovarian stromal artery Doppler indices in predicting ovarian response. J Reprod Med. 2002;47:886–90.
65. Merce L, Bau S, Bajo J. Doppler study of arterial and venous intraovarian blood flow in stimulated cycles. Ultrasound Obstet Gynecol. 2001;18:505–10.
66. Friedman CI, Danforth DR, Herbosa-Encarnacion C, Arbogast L, Alak BM, Seifer DB. Follicular fluid vascular endothelial growth factor concentrations are elevated in women of advanced reproductive age undergoing ovulation induction. Fertil Steril. 1997;68:607–12.
67. Pellicer A, Ballester MJ, Serrano MD, Mir A, Serra-Serra V, Remohi J, et al. Aetiological factors involved in the low response to gonadotrophins in infertile women with normal basal serum follicle stimulating hormone levels. Hum Reprod. 1994;9:806–11.
68. Gordon JD, Shifren JL, Foulk RA, Taylor RN, Jaffe RB. Angiogenesis in the human female reproductive tract. Obstet Gynecol Surv. 1995;50:688–97.
69. Stouffer RL, Martínez-Chequer JC, Molskness TA, Xu F, Hazzard TM. Regulation and action of angiogenic factors in the primate ovary. Arch Med Res. 2001;32:567–75.
70. Fraser HM, Duncan WC. Vascular morphogenesis in the primate ovary. Angiogenesis. 2005;8:101–16.
71. Hazzard TM, Xu F, Stouffer RL. Injection of soluble vascular endothelial growth factor receptor 1 into the preovulatory follicle disrupts ovulation and subsequent luteal function in rhesus monkeys. Biol Reprod. 2002;67:1305–12.
72. Kezele PR, Ague JM, Nilsson E, Skinner MK. Alterations in the ovarian Transcriptome during primordial follicle assembly and development 1. Biol Reprod. 2005;72:241–55.
73. Mattioli M, Barboni B, Turriani M, Galeati G, Zannoni A, Castellani G, et al. Follicle activation involves vascular endothelial growth factor production and increased blood vessel extension 1. Biol Reprod. 2001;65:1014–9.
74. Plendl J. Angiogenesis and vascular regression in the ovary. Anat Histol Embryol. 2000;29:257–66.
75. Jablonka-Shariff A, Fricke P, Grazul-Bilska A, Reynolds L, Redmer D. Size, number, cellular proliferation, and atresia of gonadotropin-induced follicles in ewes. Biol Reprod. 1994;51:531–40.
76. Berardinelli P, Russo V, Martelli A, Nardinocchi D, Di Giacinto O, Barboni B, et al. Colocalization of DNA fragmentation and caspase-3 activation during atresia in pig antral follicles. Anat Histol Embryol. 2004;33:23–7.

77. Kaczmarek MM, Schams D, Ziecik AJ. Role of vascular endothelial growth factor in ovarian physiology–an overview. Reprod Biol. 2005;5:111–36.
78. Bianco F, Basini G, Santini S, Grasselli F. Angiogenic activity of swine granulosa cells: effects of hypoxia and the role of VEGF. Vet Res Commun. 2005;29:157–9.
79. Shimizu T, Jiang J-Y, Iijima K, Miyabayashi K, Ogawa Y, Sasada H, et al. induction of follicular development by direct single injection of vascular endothelial growth factor gene fragments into the ovary of miniature gilts 1. Biol Reprod. 2003;69:1388–93.
80. Danforth DR, Arbogast LK, Ghosh S, Dickerman A, Rofagha R, Friedman CI. Vascular endothelial growth factor stimulates preantral follicle growth in the rat ovary. Biol Reprod. 2003;68:1736–41.
81. IIJIMA K, Jiang J-Y, Shimizu T, Sasada H, Sato E. Acceleration of follicular development by administration of vascular endothelial growth factor in cycling female rats. J Reprod Dev. 2005;51:161–8.
82. Wulff C, Wilson H, Wiegand SJ, Rudge JS, Fraser HM. Prevention of thecal angiogenesis, antral follicular growth, and ovulation in the primate by treatment with vascular endothelial growth factor Trap R1R2. Endocrinology. 2002;143:2797–807.
83. Murohara T, Horowitz JR, Silver M, Tsurumi Y, Chen D, Sullivan A, et al. Vascular endothelial growth factor/vascular permeability factor enhances vascular permeability via nitric oxide and prostacyclin. Circulation. 1998;97:99–107.
84. Jablonka-Shariff A, Olson LM. The role of nitric oxide in oocyte meiotic maturation and ovulation: meiotic abnormalities of endothelial nitric oxide synthase knock-out mouse oocytes 1. Endocrinology. 1998;139:2944–54.
85. Lee D-H, Joo B-S, Suh D-S, Park J-H, Choi Y-M, Lee K-S. Sodium nitroprusside treatment during the superovulation process improves ovarian response and ovarian expression of vascular endothelial growth factor in aged female mice. Fertil Steril. 2008;89:1514–21.
86. Choi K-H, Joo B-S, Sun S-T, Park M-J, Son J-B, Joo J-K, et al. Administration of visfatin during superovulation improves developmental competency of oocytes and fertility potential in aged female mice. Fertil Steril. 2012;97:1234–41. e3.
87. Ha CS, Joo BS, Kim SC, Joo JK, Kim HG, Lee KS. Estrogen administration during superovulation increases oocyte quality and expressions of vascular endothelial growth factor and nitric oxide synthase in the ovary. J Obstet Gynaecol Res. 2010;36:789–95.
88. Joo J-K, Joo B-S, Kim S-C, Choi J-R, Park S-H, Lee K-S. Role of leptin in improvement of oocyte quality by regulation of ovarian angiogenesis. Anim Reprod Sci. 2010;119:329–34.
89. Fukuhara A, Matsuda M, Nishizawa M, Segawa K, Tanaka M, Kishimoto K, et al. Visfatin: a protein secreted by visceral fat that mimics the effects of insulin. Science. 2005;307:426–30.
90. Samal B, Sun Y, Stearns G, Xie C, Suggs S, McNiece I. Cloning and characterization of the cDNA encoding a novel human pre-B-cell colony-enhancing factor. Mol Cell Biol. 1994;14:1431–7.
91. Hug C, Lodish HF. Visfatin: a new adipokine. Science. 2005;307:366–7.
92. Kim S-R, Bae S-K, Choi K-S, Park S-Y, Jun HO, Lee J-Y, et al. Visfatin promotes angiogenesis by activation of extracellular signal-regulated kinase 1/2. Biochem Biophys Res Commun. 2007;357:150–6.
93. Adya R, Tan BK, Punn A, Chen J, Randeva HS. Visfatin induces human endothelial VEGF and MMP-2/9 production via MAPK and PI3K/Akt signalling pathways: novel insights into visfatin-induced angiogenesis. Cardiovasc Res. 2008;78:356–65.
94. Xiao J, Xiao ZJ, Liu ZG, Gong HY, Yuan Q, Wang S, et al. Involvement of dimethylarginine dimethylaminohydrolase-2 in visfatin-enhanced angiogenic function of endothelial cells. Diabetes Metab Res Rev. 2009;25:242–9.
95. Bae Y-H, Bae M-K, Kim S-R, Lee JH, Wee H-J, Bae S-K. Upregulation of fibroblast growth factor-2 by visfatin that promotes endothelial angiogenesis. Biochem Biophys Res Commun. 2009;379:206–11.
96. Shen C-J, Tsai E-M, Lee J-N, Chen Y-L, Lee C-H, Chan T-F. The concentrations of visfatin in the follicular fluids of women undergoing controlled ovarian stimulation are correlated to the number of oocytes retrieved. Fertil Steril. 2010;93:1844–50.

Current Status of Ovarian Tissue Vitrification as a Fertility Preservation for the Young Cancer Patients

9

Nao Suzuki

Abstract

In young female patients, treatment of cancer can cause gonadal dysfunction, loss of fertility, and premature menopause. Cryopreservation of gametes and/or embryos and displacement or shielding of the ovaries during radiation therapy are standard methods for preserving the fertility of young female cancer patients. In 2004, Donnez et al. reported the first live birth after ovarian tissue cryopreservation and transplantation. Subsequently, ovarian tissue cryopreservation and transplantation have been applied clinically as a new approach to fertility preservation, and a new field named oncofertility has been established in Europe and the United States. It is 13 years since the first live birth was reported after ovarian tissue cryopreservation and transplantation, and this technique has now become an option that should be offered to all young female cancer patients. Cryopreservation of ovarian tissue makes it possible to preserve a larger number of primordial follicles. In addition, transplanted ovarian tissue can secrete estrogen. Therefore, this technique not only preserves fertility but may also improve symptoms of ovarian failure, prevent cardiovascular disorders related to estrogen deficiency, and lessen the decline of bone density. Currently, slow freezing is the standard method of ovarian tissue cryopreservation, while vitrification is commonly used for cryopreservation of embryos, blastocysts, oocytes, and semen. In recent years, attention has focused on use of vitrification for preservation of ovarian tissue. However, further research is needed to develop the optimum method of ovarian tissue cryopreservation and transplantation. This article reviews recent findings related to ovarian tissue vitrification.

N. Suzuki
Department of Obstetrics and Gynecology, St. Marianna University School of Medicine, Kawasaki, Japan
e-mail: nao@marianna-u.ac.jp

H. Katabuchi et al. (eds.), *Cell Biology of the Ovary*,
https://doi.org/10.1007/978-981-10-7941-2_9

Keywords
Female cancer patient · Fertility preservation · In vitro activation · Oncofertility · Ovarian tissue cryopreservation · Vitrification

9.1 Introduction

In the late 1990s, ovarian tissue cryopreservation was first employed clinically to preserve fertility in female children, adolescents, and young adults with cancer in Europe and the United States. In 2004, Donnez and colleagues reported the first successful pregnancy and live birth in a human after autotransplantation of ovarian tissue that had been cryopreserved by the slow-freezing method [1]. Ovarian tissue cryopreservation can be employed when ova cannot be collected by intravaginal procedures, when induction of ovulation is impossible in girls before menarche, and when cancer therapy must be initiated promptly and there is insufficient time to induce ovulation. Cryopreservation of ovarian tissue is performed by two methods, which are slow freezing or extremely rapid freezing (vitrification). In routine clinical treatment of infertility, the main method for cryopreservation of unfertilized ova or embryos has shifted from slow freezing to vitrification, which is currently regarded as superior in terms of both viability of thawed ova or embryos and the live birth rate [2]. It seems to be difficult to compare slow freezing with vitrification and determine which is better. The live birth rate might be thought of as a useful parameter for comparing the two procedures, but it is not actually very useful because the number of primordial follicles in ovarian tissues cannot be counted before transplantation.

9.2 Development of an Ovarian Tissue Vitrification Protocol

According to the latest report by Andersen et al. published in 2017 [3], there have been over 80 live births as result of ovarian tissue cryopreservation and transplantation, with most of the births being obtained by the slow-freezing method, excluding 3 births ([4], unpublished data). Thus, the current standard method of ovarian tissue cryopreservation is slow freezing. The major damaging factors which occur during cryopreservation are associated with chilling injury, osmotic stress, cryoprotectant toxicity, and ice crystallization (Fig. 9.1). In general, we are trying to reduce these damage by increasing cooling and warming rates during the vitrification protocol. In the past, vitrification was based on the combination of a high cooling rate and high osmotic stress. The major breakthrough in the field of vitrification came when sample volume was reduced to a level that permitted lowering of the cryoprotectant concentration. In 2006, we began to investigate a new method of ovarian tissue cryopreservation by vitrification in cynomolgus monkeys [5–7]. Initially, we performed control experiments in cynomolgus monkeys using fresh ovarian tissue (1-mm cubes of tissue), in order to establish an experimental system for ovarian

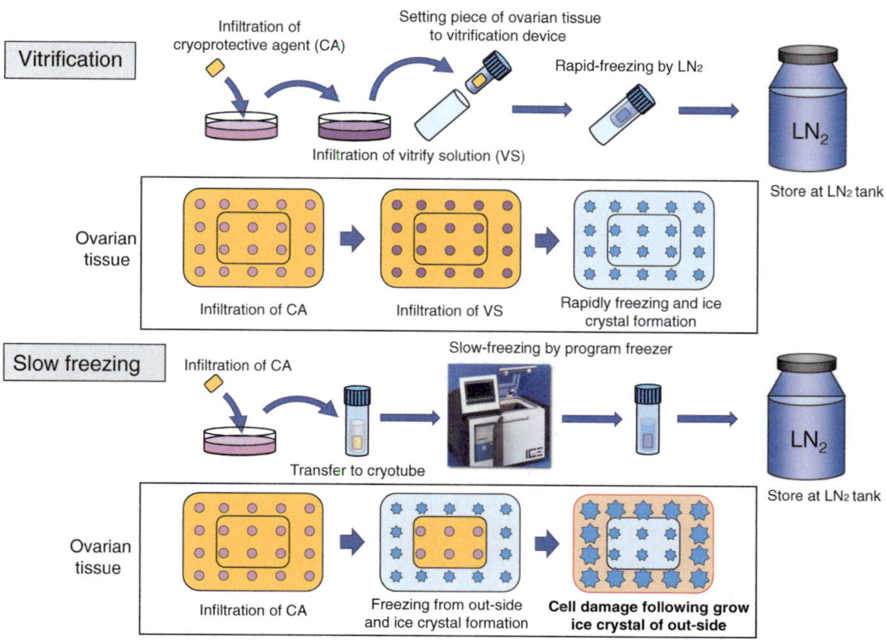

Fig. 9.1 Usually one whole ovary is removed and then cut into several pieces of ovarian tissues with the size of 1 cm in height, 1 cm in width, and 1 mm in thickness. Ovarian tissues were immediately dissected by removing medulla in medium, followed by cutting into small strips to maximize surface to mass ratios on a warm plate at 37 °C. We need to thin ovarian cortex as much as possible so that cryoprotectant immerses it well. Otherwise an ice crystal is formed and vitrification will be failed

tissue cryopreservation and transplantation. Around that time, renal subcapsular grafting of ovarian tissue in apes was reported [8]. However, we did not consider this site suitable for clinical transplantation of cryopreserved ovarian tissue, so we grafted ovarian tissue sections into the uterine serosa, mesosalpinx, peritoneum, and greater omentum. We then assessed restoration of the menstrual cycle by measuring the blood levels of follicle-stimulating hormone, luteinizing hormone, estrogen, and progesterone, and we confirmed resumption of the menstrual cycle at 64 days after transplantation based on cyclic increases and decreases of these hormones. We also performed ovarian stimulation in these monkeys and successfully harvested ova at 386 days after transplantation [6]. We concurrently attempted to develop an optimal vitrification solution, performing histologic and electron microscopic investigation of the ultrastructure of primordial follicles in cryopreserved ovarian tissue of cynomolgus monkeys. We compared VSED solution (H199 supplemented with 20% synthetic serum substitute, 3.22 mol/l ethylene glycol [EG], 2.56 mol/l dimethyl sulfoxide [DMSO], and 0.5 mol/l sucrose) [9] with VSEGP solution (H199 supplemented with 20% synthetic serum substitute, 5.64 mol/l EG, 5% polyvinylpyrrolidone, and 0.5 mol/l sucrose) [10]. When ovarian tissue was vitrified in VSEGP solution, electron microscopy revealed that primordial follicular cells formed a monolayer surrounded by flattened cells. The structure of the follicle was maintained, although there was expansion of the intercellular spaces. The organelles of the oocyte were clustered in the cytoplasm (Fig. 9.2) [5].

When ovarian tissue was vitrified in VSED solution, electron microscopy demonstrated that primordial follicular cells formed a monolayer surrounded by cuboidal cells. Most of the cytoplasmic organelles of the primordial follicular cells were found to have undergone phagocytosis (Fig. 9.2) [5]. Equilibration of ovarian tissues for 5, 10, or 20 min during vitrification was tested. Eventually, VSEGP solution with equilibration for 5 min was selected for transplantation experiments. In seven healthy female cynomolgus monkeys (aged 85–137 months), the bilateral ovaries were resected and the medulla was removed, after which the cortical tissue was subjected to cryopreservation (Fig. 9.3). Initially, 1-mm cubes of ovarian tissue were loaded into straws that were sealed to avoid exposure to liquid nitrogen (completely closed vitrification system). However, the tissue sections that fitted into the straws were very small, so it was difficult to identify the site of grafting, and ova could not be harvested successfully after ovarian tissue stimulation. Therefore, an open vitrification device that could hold ovarian tissue sections of 1 cm × 1 cm × 1 mm was developed (Ova Cryo Kit Type M) (Fig. 9.3). After vitrification and thawing, ovarian tissue was grafted into the omentum, mesosalpinx, iliac fossa, or uterine serosa. Restoration of the menstrual cycle was confirmed

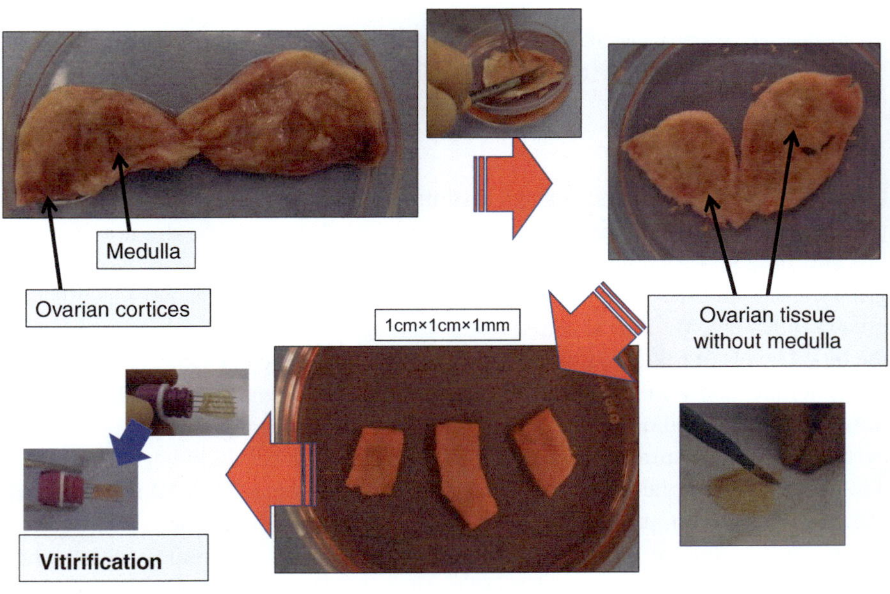

Fig. 9.2 The major damaging factors which occur during cryopreservation are associated with chilling injury, osmotic stress, cryoprotectant toxicity, and ice crystallization. In general, we are trying to reduce these damage by increasing cooling and warming rates during the vitrification protocol. In the past, vitrification was based on the combination of a high cooling rate and high osmotic stress. The major breakthrough in the field of vitrification came when sample volume was reduced to a level that permitted lowering of the cryoprotectant concentration

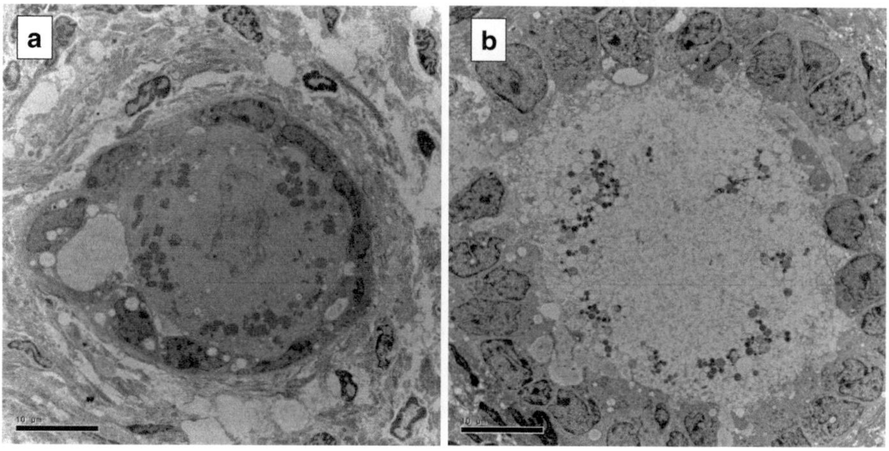

Fig. 9.3 (**a**) VSEGP: Primordial follicular cells formed a monolayer surrounded by flattened cells. The structure of the follicle was maintained, although there was expansion of the intercellular spaces. The organelles of the oocyte were clustered in the cytoplasm. (**b**) VSED: Primordial follicular cells formed a monolayer surrounded by cuboidal cells. Most of the cytoplasmic organelles of the primordial follicular cells were found to have undergone phagocytosis. Bar = 10 μm (2500×)

after growth of the grafted tissue, and we obtained good-quality embryos three times from two cynomolgus monkeys [7]. Subsequently, we started clinical application of this technology [4, 11].

9.3 Vitrification Versus Slow Freezing

The 2011 review by Amorium, a member of Donnez's group, is the most comprehensive comparison of vitrification with slow freezing for ovarian tissue cryopreservation [12]. The present review provides an updated comparison of vitrification with slow freezing.

In 2005, Lee and colleagues compared cryopreservation of simian ovaries by slow freezing or vitrification, and they reported no difference in the viability of ovarian follicles between the two methods [13]. Hovatta et al. compared propanediol-based and ethylene glycol-based cryoprotectant solutions for cryopreservation of ovarian tissue by slow freezing by evaluating ovarian follicular morphology after culture of frozen/thawed ovarian tissue for 24 h [14]. They also compared DMSO-based, propanediol-based, and ethylene glycol-based vitrification solutions at two concentrations (0.75 M and 1.5 M) each. They found no differences of the follicles between slow freezing and vitrification, while the ovarian stroma was closer to normal by electron microscopy after vitrification than slow freezing [14]. Hovatta's team subsequently developed a closed vitrification device for human ovarian tissue and obtained a favorable outcome using it [15]. In addition, Silber et al. transplanted vitrified ovarian tissue in mice and successfully obtained offspring, with their results strongly suggesting the feasibility of clinical application [16, 17]. Furthermore, Zelinski et al. investigated ovarian tissue cryopreservation and transplantation in the United States using rhesus macaques, and their electron microscopic analysis of ovarian tissue indicated that vitrification was better [18]. Subsequently, Zelinski's group developed a closed vitrification system for ovarian tissue [19].

Thus, there have been a number of reports that vitrification is superior or equivalent to slow freezing for cryopreservation of ovarian tissue. However, Amorium stated that it is difficult to directly compare slow freezing and vitrification based on the published reports [12]. He pointed out that comparison of experiments using different animal species is invalid because of structural differences of ovarian tissue among species. He also mentioned other differences among the experimental protocols, such as the composition of vitrification/cryoprotectant solutions, equilibration time, freezing system, thawing solution, incubation time after thawing, and methods of evaluation. Furthermore, Amorium suggested that evaluation of vitrification has been inadequate since there have been fewer reports about ovarian tissue cryopreservation by vitrification than slow freezing and that it is ultimately necessary to assess the safety and efficacy of procedures using high concentrations of cryoprotectants [12]. It may be considered reasonable to compare the two methods for freezing ovarian tissue on the basis of the live birth rate. However, it is impossible to count the number of primordial follicles in thawed ovarian tissue before transplantation, so even the live birth rate may not be a valid parameter for comparison of vitrification with slow freezing.

Recently, Donnez et al. mentioned that, apart from the freezing method, one of the keys to success with ovarian tissue cryopreservation and transplantation is preparation of the vascular bed before grafting [20]. They verified the importance of the vascular bed at the graft site using cynomolgus monkeys, the animals in which they first achieved live birth after ovarian tissue cryopreservation and transplantation. Their experiment involved transplantation of ovarian tissue in three groups: fresh ovarian tissue, ovarian tissue after slow freezing, and ovarian tissue after vitrification (10% DMSO, 26% ethylene glycol, 2.5% polyvinylpyrrolidone, and 1 M sucrose in MEM + Glutamax supplemented with 20 mg/ml human albumin). Each of these three groups was divided into subgroups with a healed graft bed and a freshened bed. They found that the follicular proliferation rate, percentage of antral follicles, and number of vessels per unit area were significantly higher when vitrified ovarian tissue was grafted into a freshened bed, and they concluded that this is currently the best procedure for obtaining functional ovarian tissue after autografting [20].

In 2017, Shi et al. conducted a meta-analysis of 14 studies that compared ovarian tissue cryopreservation by slow freezing and vitrification. While there was no significant difference between the two methods with regard to the percentage of intact primordial follicles, they reported that primordial follicles showed significantly less DNA fragmentation after vitrification [21]. The percentage of normal stromal cells in the ovarian tissue specimens was also significantly larger after vitrification, although there was no significant difference of primordial follicle density between the two methods. These results suggested that vitrification is possibly superior to slow freezing for cryopreservation of ovarian tissue. However, various cryoprotectants have been used for vitrification in previous studies, and this technique has not been standardized. Moreover, vitrification has a short history of clinical application, and there have been only three live births after transplantation of vitrified ovarian tissue in Japan. Thus, further validation is required before routine use of vitrification will be possible.

9.4 Clinical Application of Ovarian Tissue Cryopreservation Using Vitrification

So far, there have been a limited number of reports about the clinical application of ovarian tissue cryopreservation by vitrification. Silber and colleagues reported similar hormone production after transplantation of thawed ovarian tissue that had undergone cryopreservation by either slow freezing or vitrification, and they also found no influence of the ischemic time following transplantation of thawed ovarian tissue after cryopreservation by vitrification [22]. In 2013, Kikuchi from Silber's group reported a high collection rate of ova following the clinical application of ovarian tissue transplantation after vitrification [23].

As mentioned above, based on preclinical studies performed in cynomolgus monkeys since 2006, we have applied ovarian tissue cryopreservation by vitrification clinically since January 2010, and ovarian tissue specimens from 195 patients have undergone vitrification at our institution as of September 2015. We have also been performing ovarian tissue vitrification and transplantation combined with in vitro activation (IVA) for patients with premature ovarian insufficiency (POI), a disease associated with an early progressive decline in the number of primordial follicles. IVA was developed by Kawamura et al. using a phosphatase and tensin homolog (PTEN) inhibitor and PI3K. Ovarian tissue cryopreservation may be considered a controversial treatment for POI patients because of placing an additional burden on the patient, but we have performed it in 140 patients using vitrification for the following two reasons. First, POI is a progressive disease and the number of primordial follicles declines further over time, but cryopreservation of tissue from one ovary still allows hormone therapy to be performed in expectation of achieving follicular growth while preparing for future IVA. Second, IVA requires incubation of tissue for 48 h, and not all of the ovarian tissue harvested can be used at one time, so it is necessary to freeze the surplus ovarian tissue sections. Although it is ideal to graft ovarian tissue directly after IVA without cryopreservation, the excess ovarian tissue sections need to be cryopreserved even in this case. We have succeeded in obtaining live births by this method [4, 11]. Accordingly, ovarian tissue cryopreservation using vitrification and subsequent transplantation is thought to be promising for fertility preservation in young cancer patients, who have more primordial follicles in better condition than patients with POI. As of October 2017, we have performed ovarian tissue cryopreservation using vitrification for 87 patients with cancer or using gonadotoxic treatment. Application of ovarian tissue cryopreservation and transplantation combined with IVA should be carefully considered in cancer patients because there has been insufficient clinical experience with this procedure [24].

We are currently developing a new device for completely closed ovarian tissue vitrification and promising outcomes have been obtained, including resumption of the menstrual cycle in cynomolgus monkeys (in preparation).

Conclusion

It is anticipated that there will be further progress in fertility preservation techniques for young patients with cancer. Progress in fertility preservation technology is linked to the improved survival of young cancer patients since it increases the options for patients who wish to preserve their fertility for the future. Development of the optimum method for fertility preservation would provide young cancer patients with reassurance and allow them to concentrate on treating their primary disease. It is essential to provide patients who require gonadotoxic treatment with information on the risk of loss of fertility and the techniques for fertility preservation. Accordingly, healthcare providers need to keep up with the latest information on fertility preservation since rapid progress continues to be made in this field.

When Donnez et al. reported successful live birth after ovarian tissue cryopreservation and transplantation [1], it was a breakthrough for both fertility preservation in young cancer patients and research on reproductive medicine. We hope that ovarian tissue cryopreservation by vitrification is similarly beneficial for young cancer patients in the future.

References

1. Donnez J, Dolmans MM, Demylle D, et al. Live birth after orthotopic transplantation of cryopreserved ovarian tissue. Lancet. 2004;364:1405–10.
2. M M-B, Herrero L, Megías D, Salvanes R, Montoya MC, Cobo AC, et al. Vitrification versus slow freezing of oocytes: effects on morphologic appearance, meiotic spindle configuration, and DNA damage. Fertil Steril. 2011;95:374–7.
3. Anderson RA, Hamish W, Wallace B, Telfer EE. Ovarian tissue cryopreservation for fertility preservation: clinical and research perspectives. Hum Reprod Open. 2017;1:1–9.
4. Suzuki N, Yoshioka N, Takae S, et al. Successful fertility preservation following ovarian tissue vitrification in patients with primary ovarian insufficiency. Hum Reprod. 2015;30:608–15.
5. Hashimoto S, Suzuki N, Yamanaka M, Hosoi Y, Ishizuka B, Morimoto Y. Effects of vitrification solutions and equilibration times on the morphology of cynomolgus ovarian tissue. Reprod Biomed Online. 2010;21:501–9.
6. Igarashi S, Suzuki N, Hashimoto S, et al. Heterotopic autotransplantation of ovarian cortex in cynomolgus monkeys. Hum Cell. 2010;1:26–34.
7. Suzuki N, Hashimoto S, Igarashi S, et al. Assessment of long-term function of heterotopic transplants of vitrified ovarian tissue in cynomolgus monkeys. Hum Reprod. 2012;8:2420–9.
8. Lee DM, Yeoman RR, Battaglia DE, et al. Live birth after ovarian tissue transplant. Nature. 2004;428:137–8.
9. Ishimori H, Takahashi Y, Kanagawa H. Factors affecting survival of mouse blastocysts vitrified by a mixture of ethylene glycol and dimethyl sulfoxide. Theriogenology. 1992;38:1175–85.

10. Gandolfi F, Paffoni A, Papasso Brambilla E, et al. Efficiency of equilibrium cooling and vitrification procedures for the cryopreservation of ovarian tissue: comparative analysis between human and animal models. Fertil Steril. 2006;85:1150–6.
11. Kawamura K, Cheng Y, Suzuki N, et al. Hippo signaling disruption and Akt stimulation of ovarian follicles for infertility treatment. Proc Natl Acad Sci. 2013;43:17474–9.
12. Amorim CA, Curaba M, Van Langendonckt A, Dolmans MM, Donnez J. Vitrification as an alternative means of cryopreserving ovarian tissue. Reprod Biomed Online. 2011;23:160–86.
13. Yeoman RR, Wolf DP, Lee DM. Coculture of monkey ovarian tissue increases survival after vitrification and slow-rate freezing. Fertil Steril. 2005;83:1248–54.
14. Keros V, Xella S, Hultenby K, et al. Vitrification versus controlled-rate freezing in cryopreservation of human ovarian tissue. Hum Reprod. 2009;7:1670–83.
15. Sheikhi M, Hultendby K, Niklasson B, Lundqvist M, Hobatta O. Clinical grade vitrification: an ultrastructural analysis of follicles and stroma. Hum Reprod. 2011;26:594–603.
16. Kagawa N, Kuwayama M, Nakata K, et al. Production of the first offspring from oocytes derived from fresh and cryopreserved pre-antral follicles of adult mice. Reprod Biomed Online. 2007;14:693–9.
17. Kagawa N, Silber S, Kuwayama M. Successful vitrification of bovine and human ovarian tissue. Reprod Biomed Online. 2009;18:568–77.
18. Ting AY, Yeoman RR, Lawson S, Zelinski MB. In vitro development of secondary follicles from cryopreserved Rhesus macaque ovarian tissue after slow-rate-freeze or vitrification. Hum Reprod. 2011;26:2461–72.
19. Ting AY, Yeoman RR, Campos JR, et al. Morphological and functional preservation of preantral follicles after vitrification of macaque ovarian tissue in a closed system. Hum Reprod. 2013;28:1267–79.
20. Dolmans MM, Binda MM, Jacobs S, et al. Impact of the cryopreservation technique and vascular bed on ovarian tissue transplantation in cynomolgus monkeys. J Assist Reprod Genet. 2015;32:1251–62.
21. Silber S, Kagawa N, Kuwayama M, Gosden R. Duration of fertility after fresh and frozen ovary transplantation. Fertil Steril. 2010;94:2191–6.
22. Shi Q, Xie Y, Wang Y, Li S. Vitrification versus slow freezing for human ovarian tissue cryopreservation: a systematic review and meta-analysis. Sci Rep. 2017;7:8538.
23. Kikuchi I, Kagawa N, Silber S, et al. Oophorectomy for fertility preservation via reduced-port laparoscopic surgery. Surg Innov. 2013;20:219–24.
24. Suzuki N. Ovarian tissue cryopreservation using vitrification and/or in vitro activated technology. Hum Reprod. 2015;11:2461–2.

Molecular Pathogenesis of Ovarian Cancer: An Inextricable Maze

10

Masaki Mandai

Abstract

We have long been believed that the origin of ovarian cancer is the ovarian surface epithelium. Many clinical and biological findings support this hypothesis, but there are also some controversies. Recently completely new idea has emerged in which tubal epithelium is considered as an origin of high-grade serous ovarian cancer. This hypothesis is so sophisticated as to address many controversies that cannot be explained by the ovarian surface epithelium hypothesis. However, final conclusions on the true picture of ovarian carcinogenesis have not been obtained yet.

Specific types of ovarian cancer, including endometrioid and clear cell carcinoma, are known to arise in endometriotic cyst. While endometrioid carcinoma may develop under the influence of estrogen stimulation, clear cell carcinoma may develop as the result of stressful environment in endometriotic cyst. Characteristic glucose metabolism of clear cell carcinoma may be associated with stress response and resistance to chemotherapy.

Keywords

Ovarian surface epithelium · STIC · Endometriotic cyst · Oxidative stress · Glucose metabolism

M. Mandai
Department of Gynecology and Obstetrics, Kyoto University Graduate School of Medicine, Kyoto, Japan
e-mail: mandai@kuhp.kyoto-u.ac.jp

H. Katabuchi et al. (eds.), *Cell Biology of the Ovary*,
https://doi.org/10.1007/978-981-10-7941-2_10

10.1 Introduction

Recent efforts to explore the molecular pathogenesis of ovarian cancer have made great progress. However, these advances have provided more confusion rather than solutions, and it seems as if we are now trapped in a mysterious maze. However, if we look back the history of research on the origin of ovarian cancer, we find that there have also been many mysteries and controversies from the beginning.

Since I started my career as a gynecologic oncologist some 25 years ago, the origin of ovarian cancer has continued to be a great mystery. Fundamentally, the ovary, an organ specific for reproduction, has two separate functions: to grow eggs and to secrete hormones. However, the ovary has no function of exocrine secretion. In fact, there is no apparent glandular tissue in the ovary. Nevertheless, more than 90% of ovarian malignant tumors are epithelial tumors. From where and how do these tumors arise? A well-accepted dogma for ovarian carcinogenesis is ovarian surface inclusion theory, which proposes that the origin of epithelial tumors is the featureless line of cells covering the ovarian surface called the "ovarian surface epithelium" or OSE. These cells are actually a continuation of the peritoneal surface and of mesothelial rather than epithelial origin. However, for some reason, OSE cells change to various types of epithelium in the process of transformation and serve as the origin of ovarian cancer [1–3] (Fig. 10.1).

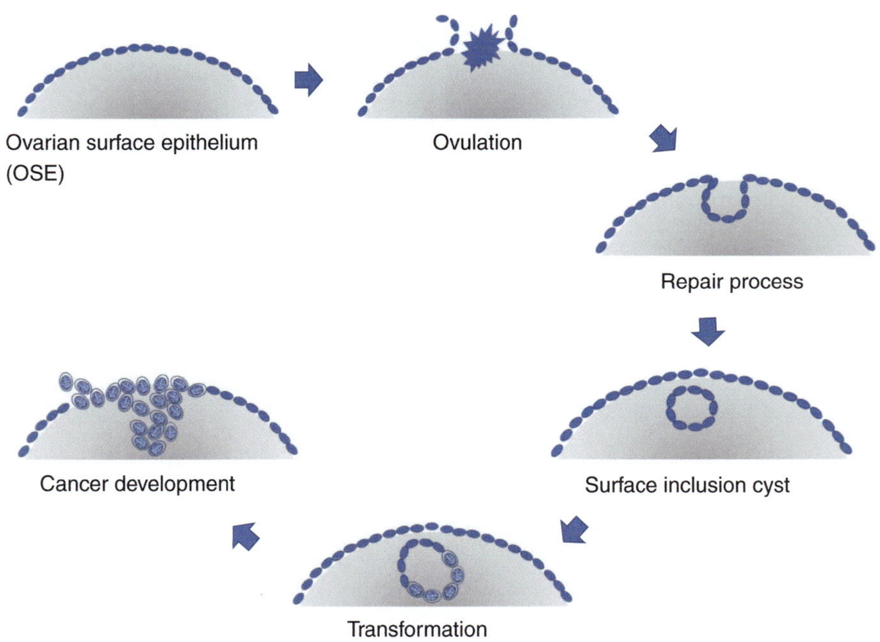

Fig. 10.1 Previous understanding of the origin of ovarian cancer

Abundant indirect evidence supports this hypothesis. For example, we often observe cystic structures within the ovary, which are designated as surface inclusion cysts because it is believed that they are the result of invagination of the OSE after rupture of the ovarian surface at the time of ovulation. The cells of the cystic wall show various epithelial morphological features, which were attributed to the transformation process to the epithelial tumor [3–5] (Fig. 10.1). Clinical and epidemiological evidence that the number of lifetime ovulations of an individual woman is associated with the risk of ovarian cancer development and that the use of oral contraceptives decreases the risk of ovarian cancer, possibly by decreasing the number of ovulations, are consistent with this hypothesis [6, 7]. We have shown that this hypothesis is correct in at least some mucinous ovarian carcinomas. We observed a stepwise occurrence of K-ras gene mutation in accordance with the morphological transformation of the mucinous epithelium from benign to carcinoma [8]. These findings indicate that in the carcinogenic process of these cancers, molecular events accumulate step by step as the inclusion epithelium transforms to a benign tumor and further to a malignant tumor.

10.2 Ovarian Surface Epithelium as the Putative Origin of Serous Ovarian Cancer

Despite significant indirect evidence, there was no direct evidence that OSE is the origin of ovarian cancer. Therefore, we attempted to show this carcinogenic process directly in vivo. We generated a primary culture of OSE cells and showed that their growth was affected by hormonal treatment [9]. Then we attempted to transform OSE into cancer by transfecting cultured human OSE cells with oncogenes. Immortalization of OSE cells was achieved by the transfection of SV40 large T antigen (LT) and human telomerase reverse transcriptase (hTERT); however, the resultant cells exhibited no tumorigenesis. Additional transfection of either c-erbB-2 or mutant Ha-ras into the immortalized cells resulted in anchorage-independent growth and tumorigenesis in mice with incidences of 50% and 40%, respectively [10].

However, unfortunately for us, we were not the first to report these observations. Professor Katabuchi and Professor Tashiro's group reported similar results 2 years before us [11], and more importantly, they subsequently generated immortalized OSE cells with a normal karyotype [12]. This cell line has since served as the standard cell line representing a normal counterpart of ovarian cancer cell lines. These data indicated that OSE can be transformed into a malignant tumor via multiple genetic events. Thus, Japanese gynecologists competed with each other, took initiatives, and made significant contributions in the field of ovarian carcinogenesis. I personally think the OSE hypothesis remains valid and cannot be completely declined. However, a completely new idea has recently emerged.

10.3 Fallopian Tube as a Putative Origin of Serous Ovarian Cancer

A recent trend in ovarian carcinogenesis research is the hypothesis that at least a significant proportion of high-grade serous carcinoma is of tubal origin, as proposed by Professor Kurman's group [13]. They propose that tubal epithelia somehow migrate to the ovarian surface and transform into carcinoma. Careful examination of the fallopian tubes reveals that a significant proportion of ovarian cancer cases also have abnormal lesions in the ovary, designated as STIC (serous tubal intraepithelial carcinoma). This tubal origin hypothesis is very attractive and sophisticated and can answer many controversies of the OSE hypothesis, including the lack of epithelium in the ovary. However, a careful review of the paper published by Professor Kurman's group reveals that this hypothesis also contains many controversies. The debate between these two hypotheses on the origin of serous ovarian cancer is ongoing. For example, an article by Professor Kurman's group suggested that some surface inclusion cysts are actually of tubal origin [14]. Professor Auersperg, a proponent of the OSE hypothesis, stated that there appears to be no evidence supporting Professor Kurman's proposal [15], receiving a negative response from Professor Kurman [16]. It is surprising to witness these famous professors engaging in such a frank discussion, and it appears that both hypotheses have advantages and defects. As Professor Auersperg proposed, the origin of serous ovarian cancer may be somewhere else [17]. I hope that young researchers from Japan will participate in this dispute from their original points of view and not simply accept either idea but contribute to solving this longstanding issue.

Another aspect of this dispute is that Professor Auersperg's group mostly uses rather conventional methods, such as immunohistochemistry, whereas Professor Kurman's group uses more modern methods very effectively, namely, comprehensive molecular analysis. It is obvious that this type of molecular approach, if used effectively, is a powerful means of solving old pathological disputes in a very sophisticated manner. In fact, Professor Kurman and Professor Shih's group revisited and provided new insights on old pathological issues such as the teratomatous origin of mucinous ovarian tumors and the association between Brenner tumors and mucinous tumors by employing molecular analysis. It has long been believed that a subset of primary ovarian mucinous tumors is derived from mature teratomas. Professor Kurman's group performed microsatellite genotyping using short tandem repeat markers using eight mucinous tumors associated with a teratoma and 22 ovarian mucinous carcinomas without a teratoma to determine if they were clonally related. Most of the mucinous tumors associated with a teratoma showed a high or complete degree of allelotype matching that differed from the somatic allelotypes of the normal control tissue. By contrast, none of the ovarian mucinous carcinomas without a teratoma showed pure homozygosity. They concluded that most mucinous tumors associated with a teratoma are derived from the teratoma, whereas the majority of pure mucinous tumors in young women in whom a teratoma is not present are not derived from a teratoma [18]. Similarly, it has been proposed that some intestinal-type mucinous tumors may be derived from Brenner tumors, which they frequently coincide with. Professor Kurman's group performed a clonality analysis of the two components of ten combined Brenner and mucinous tumors and identified a shared clonal relationship between the two different tumor components. These observations suggest that some pure mucinous tumors may develop from a Brenner tumor in which the Brenner tumor component becomes compressed and obliterated by an expanding mucinous neoplasm [19]. Professor Kurman himself stated the importance of including these analyses in investigations of the origin of ovarian cancer [20], and very recently, two reports of comprehensive genetic analyses of this issue have been published. A group with Professor Kurman and Professor Shih performed whole-exome sequencing and copy number analyses of laser capture microdissected fallopian tube lesions, ovarian cancers, and metastases from nine patients. The majority of tumor-specific alterations in ovarian cancers were present in STICs. Evolutionary analyses reveal that p53 signatures and STIC lesions are precursors of ovarian carcinoma and that the interval between the development of a STIC and the initiation of ovarian carcinoma may be up to 7 years [21]. Another group also conducted a multicenter integrated genomic analysis of advanced-stage tumors with and without STIC lesions and normal tissues. Their data suggested that the molecular features of high-grade serous ovarian cancer with and without associated STIC lesions are mostly shared, indicating a common biological origin [22]. Thus, recent molecular analyses generally suggest that at least STIC lesion and high-grade serous ovarian cancer have shared genetic features and may have common origin. Nevertheless, I feel that the true picture of serous ovarian carcinogenesis remains unclear, with many other possibilities [17]. We should be aware that it is a dangerous attitude to simply accept these ideas without any criticism, even though they are very attractive hypotheses. We should judge these ideas by our own data and from own viewpoint.

10.4 Ovarian Cancer Associated with Ovarian Endometriotic Cyst

Another new idea on ovarian carcinogenesis is the relationship between ovarian cancer and ovarian endometriosis. A significantly higher incidence of malignant transformation is observed in ovarian endometriotic cysts compared with other benign ovarian tumors such as serous and mucinous ovarian cysts [23, 24], but the underlying reasons have been unclear. Another feature of malignant transformation of endometriotic cysts is that two specific histologic types of ovarian carcinoma, endometrioid and clear cell carcinomas, arise almost exclusively from endometriotic cysts. Serous cancer, the most common type of ovarian cancer, rarely develops [24]. Again, the reasons for these discrepancies are unclear.

The carcinogenic process of endometrioid carcinoma is relatively easy to assume because we have a similar model of carcinogenesis in the uterus, that is, endometrial cancer (Fig. 10.2). In the uterine endometrium, endometrioid carcinoma usually develops via a stepwise carcinogenic process from normal endometrium to endometrial hyperplasia and carcinoma. A similar stepwise carcinogenesis from

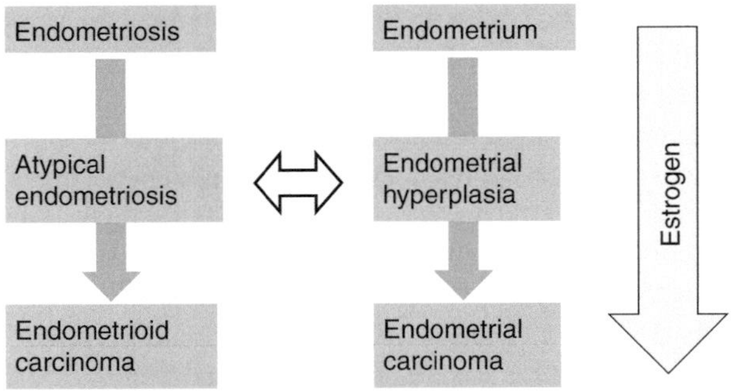

Fig. 10.2 Possible developmental procedure of endometrioid carcinoma from endometriosis2

endometriosis to endometrioid carcinoma is suggested pathologically and by molecular analyses. Pathologically, we occasionally encounter so-called atypical endometriosis, a putative intermediate lesion between benign endometriosis and ovarian cancer. In a review of 194 cases of ovarian endometriosis, Czernobilsky and Morris observed severe cytological atypia in 3.6% and adenomatous, hyperplasia-like lesions in 2.0% [25]. LaGrenade and Silverberg presented four cases of ovarian carcinomas contiguous with atypical glandular epithelial changes in endometriosis [26]. Fukunaga et al. reported that as many as 61% of cases of endometriosis coexisting with ovarian cancer had atypical endometriosis foci [27], and Ogawa et al. reported atypical endometriosis in 78% of the cases with endometriosis-associated ovarian cancer [28]. Several studies also have revealed an intermediate nature of atypical endometriosis by molecular analyses. Sáinz de la Cuesta et al. evaluated the immunohistochemical expression of p53 in normal endometrium, endometriosis, atypical endometriosis, and ovarian cancer associated with endometriosis and found that 82.4% of cancers associated with endometriosis and all atypical endometriosis samples showed P53 overexpression, whereas only 11.8% of the endometriosis samples and none of the endometrium samples showed P53 overexpression [29]. In a report by Obata et al., frequent loss of heterozygosity (LOH) was observed on chromosome 6q (60.0%) and chromosome 10q (40.0%) in ovarian atypical endometriosis [30]. These pathological observations suggest that, in some cases, ovarian cancer develops from an endometriotic cyst via atypical endometriosis, a premalignant lesion. As endometrioid carcinoma in the endometrium develops under the influence of estrogen, it seems natural to assume that endometrioid carcinoma arising from an endometriotic cyst also affected by estrogen via a similar carcinogenic mechanism.

10.5 Stressful Microenvironment in Endometriotic Cysts and Carcinogenesis

By contrast, there is apparently no clear explanation of why clear cell carcinoma often develops from endometriotic cysts. Moreover, why is clear cell carcinoma a chemo-resistant cancer. We focused on the characteristic microenvironment as a major driving force of the development of this cancer (Fig. 10.3). Inside of an

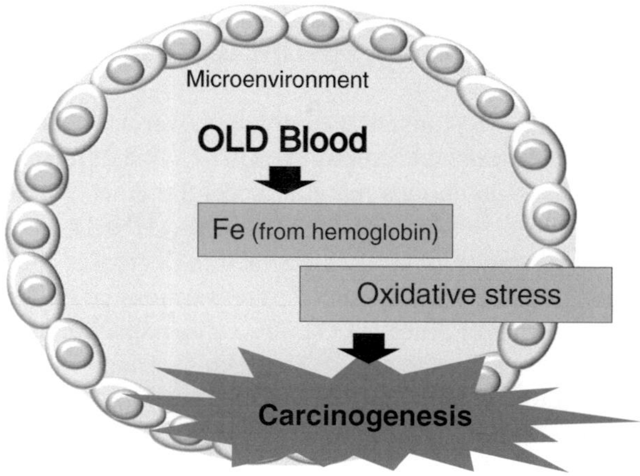

Fig. 10.3 Microenvironment in endometriotic cyst may contribute malignant transformation

endometriotic cyst, surface epithelial cells are continuously exposed to old blood as the result of repeated bleeding, and this fluid may contain large amounts of iron derived from blood hemoglobin. We measured the free-iron concentration within the ovarian cyst and found that the fluid in chocolate cysts contained extremely high levels of free iron compared with other epithelial cysts [31]. Iron is a well-known inducer of oxidative stress, and oxidative stress may lead to carcinogenesis [32]. Therefore, we measured several markers of oxidative stress in endometriotic cysts. The fluid in endometriotic cysts had significantly higher oxidative stress, cellular, and DNA damage. We also examined whether this fluid can cause DNA damage. Culturing V79 cells with endometriotic cyst fluid induced significantly more DNA damage [31]. These data suggest that the content of chocolate cyst fluid, especially free iron, causes DNA damage via severe oxidative stress.

Then, why clear cell carcinoma, a rare type of ovarian cancer, arises in an endometriotic cyst? Is this development also explained by the microenvironment? To address this issue, we used microarray analyses and extracted gene signatures distinguishing ovarian clear cell carcinoma from other ovarian cancers [33]. A signature comprising 320 genes were developed and designated as the OCCC signature. Next, we analyzed what types of genes the OCCC signature contains and found that it contains mostly stress-response genes and glucose metabolism-related genes. Interestingly, when we cultured OSE cells and added endometriotic cyst fluid, time-dependent induction of the OCCC signature was observed, suggesting that the expression of these genes specific to clear cell carcinoma is primarily induced by the microenvironment in chocolate cysts [33]. Thus, we came to a unique hypothesis that the stressful microenvironment in chocolate cysts not only promotes carcinogenesis but also contributes to the occurrence of a specific histological type of cancer, namely, clear cell carcinoma [34]. In addition, by analyzing the genes included in the OCCC signature, it became clear that there were two keywords, stress response and glucose metabolism. Therefore, the next question was whether this metabolism is related to the biology of clear cell carcinoma.

10.6 Ovarian Clear Cell Carcinoma as a Stress-Resistant Cancer with Specific Glucose Metabolism

To further investigate the biology of ovarian clear cell carcinoma, we searched for the key molecule among the OCCC signature genes. We ultimately identified HNF1b as potentially playing a central role [33–36]. Therefore, we knocked down the expression of HNF1b in the HNF1b-high OCCC cell line RMG-2. Downregulation of HNF1b resulted in decreased glucose uptake and glycolysis, indicating that in ovarian clear cell carcinoma, HNF1b promotes the uptake and consumption of glucose [37]. The cancer-specific increase in glycolysis is known as the Warburg effect. In 1956, Otto Warburg suggested that cancer cells predominantly obtain energy by glycolysis, whereas the majority of normal cells produce energy by oxidative phosphorylation in mitochondria [38]. In other words, cancer cells gain energy in the same way that athletes gain energy while running at high speed. But what is the purpose of this metabolism? We examined whether the stress resistance of cancer cells is associated with HNF1b expression. By comparing intracellular stress, stress resistance, and comprehensive metabolome assays including glucose metabolism between HNF1b-high- and HNF1b-low-expression RMG2 clear cell carcinoma cells, we found that HNF1b-high cancer cells are more stress resistant [39]. They were also resistant to chemotherapy, another form of cellular stress. Importantly, this stress resistance was observed only when CCC cells had a sufficient supply of glucose. When glucose was removed from the medium, HNF1b-high cells no longer exhibited stress resistance. These results clearly suggest that ovarian clear cell carcinoma overexpresses HNF1b in response to the stressful microenvironment in endometriotic cysts and that HNF1b contributes to the acquisition of stress resistance (as well as chemoresistance) by increasing specific glucose metabolism, namely, the Warburg effect [40]. Thus, we hypothesize that the stressful microenvironment is the primary cause of the development of chemoresistant cancer in endometriotic cysts (Fig. 10.4).

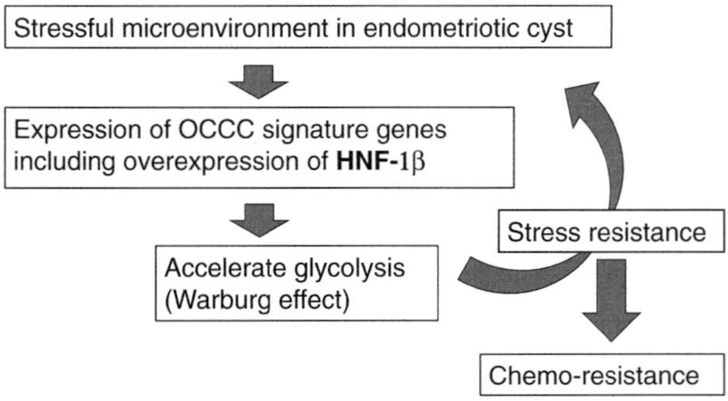

Fig. 10.4 Stress response and chemo-resistance of ovarian clear cell carcinoma

10.7 Ovarian Carcinogenesis: What We Will Know in the Future

Finally, I will discuss future directions of the molecular understanding of ovarian cancer biology. Recent ideas have completely altered our classic understanding of ovarian carcinogenesis. First, most serous ovarian cancer originates from the fallopian tube. It has also been suggested that the vast majority of mucinous ovarian cancers are actually metastases from digestive organs. Endometrioid and clear cell cancer of the ovary arise from the endometriosis, which is believed to be a transplantation of the endometrium. If these hypotheses are correct, it becomes that there is no genuine ovarian cancer that originates from the ovary and the OSE hypothesis that we believed in for many years is false. As I mentioned earlier, these ideas remain hypotheses and could still be reconsidered, as in the case of the OSE hypothesis (Fig. 10.5).

However, a more drastic change in our understanding of the biology of ovarian cancer has emerged from recent research on ovarian cancer. In 2011, TCGA analysis of high-grade serous ovarian cancer indicated that the pathologically single category of high-grade serous cancer can be subdivided into four groups, namely, proliferative, differentiated, mesenchymal, and immunoreactive, based on comprehensive gene expression analysis [41]. We were interested in whether this type of classification reflects pathological findings. We intermingled our own cases of serous cancer with the TCGA data and performed re-clustering to determine which

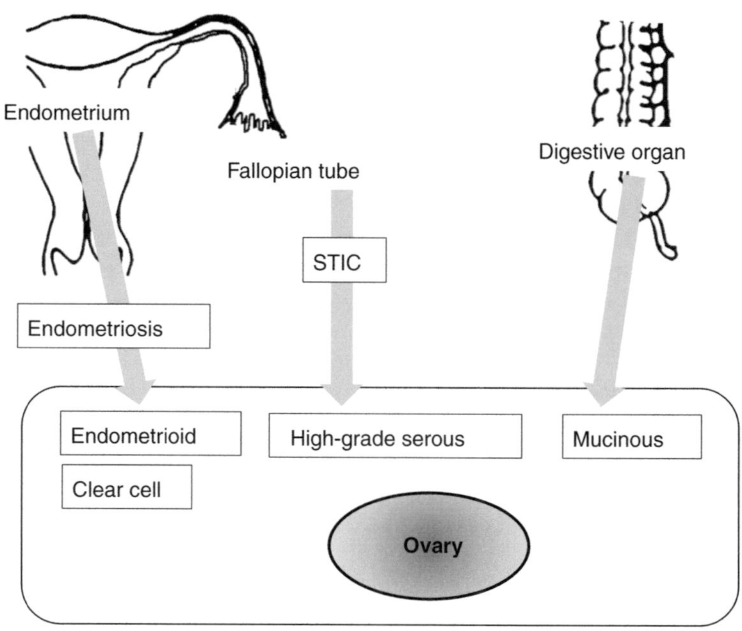

Fig. 10.5 Recent understanding of the origin of ovarian cancer

subcategories our own cases belonged to. We then examined these specimens under the microscope [42]. The results were interesting because molecular subgrouping did reflect the pathological findings. The cases that belonged to the proliferative subtype showed a relatively solid, poorly differentiated phenotype, whereas the cases belonging to the differentiated subtype showed more differentiated, papillary growth. The mesenchymal subtype showed pathologic features of isolated cancer cells invading the stroma and causing a destructive stromal reaction. Finally, the immunoreactive subtype showed massive infiltration of immune cells into the tumor site. The most important point here is that this classification is associated with patient prognosis. Among the four groups, the mesenchymal type has the worst prognosis, whereas the immunoreactive type has the best, with no notable difference in survival between the proliferative and differentiated subtypes. Pathology has focused on the morphology of cancer itself. Pathological classification is solely based on the appearance of cancer lesions. However, comprehensive molecular analyses have clearly shown that the most important factors affecting the patient are not the appearance of the cancer itself but its surroundings, such as stromal cells or immune cells. In other words, the microenvironment is more important and reflects cancer biology. We recently proposed a novel pathological classification of high-grade serous ovarian cancer reflecting the TCGA findings that includes microenvironment factors. Obviously, our methods of understanding and defining cancer are becoming increasingly complicated with the rapid advances in our knowledge, especially in the field of genetic analysis. Therefore, in terms of understanding the true biological nature of each cancer, it is now insufficient to classify cancers only by pathology. To define each individual cancer, we should integrate all types of factors, including pathological findings, gene expression, genetic mutations, metabolism, and, most importantly, microenvironment. Ultimately, it will be meaningless to classify each cancer. We will treat them individually without any classification, possibly through the development of technology for handling massive information using IT technology (artificial intelligence). The use of such technologies may allow us to understand the more individualized and detailed mechanisms of ovarian cancer pathogenesis in each case.

In summary, ideas on ovarian carcinogenesis have changed significantly over time, mainly due to advances in analytic tools: simple pathology, immunohistochemistry, molecular analysis, comprehensive expression analysis, and whole-genome analysis. Nevertheless, the real picture of the pathogenesis of ovarian cancer still appears to be an inextricable maze awaiting future research development.

References

1. Feeley KM, Wells M. Precursor lesions of ovarian epithelial malignancy. Histopathology. 2001;38:87–95. Review
2. Scully RE. Pathology of ovarian cancer precursors. J Cell Biochem Suppl. 1995;23:208–18. Review
3. Wang DP, Konishi I, Koshiyama M, Nanbu Y, Iwai T, Nonogaki H, Mori T, Fujii S. Immunohistochemical localization of c-erbB-2 protein and epidermal growth factor recep-

tor in normal surface epithelium, surface inclusion cysts, and common epithelial tumours of the ovary. Virchows Arch A Pathol Anat Histopathol. 1992;421:393–400.
4. Hutson R, Ramsdale J, Wells M. p53 protein expression in putative precursor lesions of epithelial ovarian cancer. Histopathology. 1995;27:367–71.
5. Aoki Y, Kawada N, Tanaka K. Early form of ovarian cancer originating in inclusion cysts. A case report. J Reprod Med. 2000;45:159–61.
6. Casagrande JT, Louie EW, Pike MC, Roy S, Ross RK, Henderson BE. "Incessant ovulation" and ovarian cancer. Lancet. 1979;2:170–3.
7. Booth M, Beral V, Smith P. Risk factors for ovarian cancer: a case–control study. Br J Cancer. 1989;60:592–8.
8. Mandai M, Konishi I, Kuroda H, Komatsu T, Yamamoto S, Nanbu K, Matsushita K, Fukumoto M, Yamabe H, Mori T. Heterogeneous distribution of K-ras-mutated epithelia in mucinous ovarian tumors with special reference to histopathology. Hum Pathol. 1998;29:34–40.
9. Kuroda H, Mandai M, Konishi I, Tsuruta Y, Kusakari T, Kariya M, Fujii S. Human ovarian surface epithelial (OSE) cells express LH/hCG receptors, and hCG inhibits apoptosis of OSE cells via up-regulation of insulin-like growth factor-1. Int J Cancer. 2001;91:309–15.
10. Kusakari T, Kariya M, Mandai M, Tsuruta Y, Hamid AA, Fukuhara K, Nanbu K, Takakura K, Fujii S. C-erbB-2 or mutant Ha-ras induced malignant transformation of immortalized human ovarian surface epithelial cells in vitro. Br J Cancer. 2003;89:2293–8.
11. Nitta M, Katabuchi H, Ohtake H, Tashiro H, Yamaizumi M, Okamura H. Characterization and tumorigenicity of human ovarian surface epithelial cells immortalized by SV40 large T antigen. Gynecol Oncol. 2001;81:10–7.
12. Maeda T, Tashiro H, Katabuchi H, Begum M, Ohtake H, Kiyono T, Okamura H. Establishment of an immortalised human ovarian surface epithelial cell line without chromosomal instability. Br J Cancer. 2005;93:116–23.
13. Kurman RJ. Origin molecular pathogenesis of ovarian high-grade serous carcinoma. Ann Oncol. 2013;24:16–21.
14. Banet N, Kurman RJ. Two types of ovarian cortical inclusion cysts: proposed origin and possible role in ovarian serous carcinogenesis. Int J Gynecol Pathol. 2015;34:3–8.
15. Auersperg N. Article by Natalie Banet and Robert J. Kurman: Two types of ovarian cortical inclusion cysts: proposed origin and possible role in ovarian serous carcinogenesis. Int. J. Gynecol. Pathol. 2015;34:3-8. Int J Gynecol Pathol. 2015;34:303–4.
16. Kurman RJ, Banet N. Re: Article by Natalie Banet and Robert J. Kurman: Two types of ovarian cortical inclusion cysts: proposed origin and possible role in ovarian serous carcinogenesis; Int. J. Gynecol. Pathol. 2015;34:3-8. Int J Gynecol Pathol. 2015;34:304–5.
17. Auersperg N. The origin of ovarian carcinomas: a unifying hypothesis. Int J Gynecol Pathol. 2011;30:12–21.
18. Wang Y, Schwartz LE, Anderson D, Lin MT, Haley L, Wu RC, Vang R, Shih IM, Kurman RJ. Molecular analysis of ovarian mucinous carcinoma reveals different cell of origins. Oncotarget. 2015;6:22949–58.
19. Wang Y, Wu RC, Shwartz LE, Haley L, Lin MT, Shih IM, Kurman RJ. Clonality analysis of combined Brenner and mucinous tumours of the ovary reveals their monoclonal origin. J Pathol. 2015;237:146–51.
20. Kurman RJ, Shih IM. The Dualistic Model of ovarian carcinogenesis: revisited, revised, and expanded. Am J Pathol. 2016;186:733–47.
21. Labidi-Galy SI, Papp E, Hallberg D, Niknafs N, Adleff V, Noe M, Bhattacharya R, Novak M, Jones S, Phallen J, Hruban CA, Hirsch MS, Lin DI, Schwartz L, Maire CL, Tille JC, Bowden M, Ayhan A, Wood LD, Scharpf RB, Kurman R, Wang TL, Shih IM, Karchin R, Drapkin R, Velculescu VE. High grade serous ovarian carcinomas originate in the fallopian tube. Nat Commun. 2017;8:1093.
22. Ducie J, Dao F, Considine M, Olvera N, Shaw PA, Kurman RJ, Shih IM, Soslow RA, Cope L, Levine DA. Molecular analysis of high-grade serous ovarian carcinoma with and without associated serous tubal intra-epithelial carcinoma. Nat Commun. 2017;8:990.

23. Kobayashi H, Sumimoto K, Kitanaka T, Yamada Y, Sado T, Sakata M, Yoshida S, Kawaguchi R, Kanayama S, Shigetomi H, Haruta S, Tsuji Y, Ueda S, Terao T. Ovarian endometrioma—risks factors of ovarian cancer development. Eur J Obstet Gynecol Reprod Biol. 2008;138:187–93.
24. Mandai M, Yamaguchi K, Matsumura N, Baba T, Konishi I. Ovarian cancer in endometriosis: molecular biology, pathology, and clinical management. Int J Clin Oncol. 2009;14:383–91.
25. Czernobilsky B, Morris WJ. A histologic study of ovarian endometriosis with emphasis on hyperplastic and atypical changes. Obstet Gynecol. 1979;53:318–23.
26. LaGrenade A, Silverberg SG. Ovarian tumors associated with atypical endometriosis. Hum Pathol. 1988;19:1080–4.
27. Fukunaga M, Nomura K, Ishikawa E, Ushigome S. Ovarian atypical endometriosis: its close association with malignant epithelial tumours. Histopathology. 1997;30:249–55.
28. Ogawa S, Kaku T, Amada S, Kobayashi H, Hirakawa T, Ariyoshi K, Kamura T, Nakano H. Ovarian endometriosis associated with ovarian carcinoma: a clinicopathological and immunohistochemical study. Gynecol Oncol. 2000;77:298–304.
29. Sáinz-de-la-Cuesta R, Izquierdo M, Cañamero M, Granizo JJ, Manzarbeitia F. Increased prevalence of p53 overexpression from typical endometriosis to atypical endometriosis and ovarian cancer associated with endometriosis. Eur J Obstet Gynecol Reprod Biol. 2004;113:87–93.
30. Obata K, Hoshiai H. Common genetic changes between endometriosis and ovarian cancer. Gynecol Obstet Investig. 2000;50:39–43.
31. Yamaguchi K, Mandai M, Toyokuni S, Hamanishi J, Higuchi T, Takakura K, Fujii S. Contents of endometriotic cysts, especially the high concentration of free iron, are a possible cause of carcinogenesis in the cysts through the iron-induced persistent oxidative stress. Clin Cancer Res. 2008;14:32–40.
32. Lee JC, Son YO, Pratheeshkumar P, Shi X. Oxidative stress and metal carcinogenesis. Free Radic Biol Med. 2012;53:742–57.
33. Yamaguchi K, Mandai M, Oura T, Matsumura N, Hamanishi J, Baba T, Matsui S, Murphy SK, Konishi I. Identification of an ovarian clear cell carcinoma gene signature that reflects inherent disease biology and the carcinogenic processes. Oncogene. 2010;29:1741–52.
34. Mandai M, Matsumura N, Baba T, Yamaguchi K, Hamanishi J, Konishi I. Ovarian clear cell carcinoma as a stress-responsive cancer: influence of the microenvironment on the carcinogenesis and cancer phenotype. Cancer Lett. 2011;310:129–33.
35. Yamaguchi K, Huang Z, Matsumura N, Mandai M, Okamoto T, Baba T, Konishi I, Berchuck A, Murphy SK. Epigenetic determinants of ovarian clear cell carcinoma biology. Int J Cancer. 2014;135:585–97.
36. Yamaguchi K, Matsumura N, Mandai M, Baba T, Konishi I, Murphy SK. Epigenetic and genetic dispositions of ovarian carcinomas. Oncoscience. 2014;1:574–9.
37. Okamoto T, Mandai M, Matsumura N, Yamaguchi K, Kondoh H, Amano Y, Baba T, Hamanishi J, Abiko K, Kosaka K, Murphy SK, Mori S, Konishi I. Hepatocyte nuclear factor-1β (HNF-1β) promotes glucose uptake and glycolytic activity in ovarian clear cell carcinoma. Mol Carcinog. 2015;54:35–49.
38. Warburg O. On the origin of cancer cells. Science. 1956;123:309–14.
39. Amano Y, Mandai M, Yamaguchi K, Matsumura N, Kharma B, Baba T, Abiko K, Hamanishi J, Yoshioka Y, Konishi I. Metabolic alterations caused by HNF1β expression in ovarian clear cell carcinoma contribute to cell survival. Oncotarget. 2015;6:26002–17.
40. Mandai M, Amano Y, Yamaguchi K, Matsumura N, Baba T, Konishi I. Ovarian clear cell carcinoma meets metabolism; HNF-1β confers survival benefits through the Warburg effect and ROS reduction. Oncotarget. 2015;6:30704–14.
41. Cancer Genome Atlas Research Network. Integrated genomic analyses of ovarian carcinoma. Nature. 2011;474:609–15.
42. Murakami R, Matsumura N, Mandai M, Yoshihara K, Tanabe H, Nakai H, Yamanoi K, Abiko K, Yoshioka Y, Hamanishi J, Yamaguchi K, Baba T, Koshiyama M, Enomoto T, Okamoto A, Murphy SK, Mori S, Mikami Y, Minamiguchi S, Konishi I. Establishment of a novel histopathological classification of high-grade serous ovarian carcinoma correlated with prognostically distinct gene expression subtypes. Am J Pathol. 2016;186:1103–13.

Clinical Management of Epithelial Ovarian Cancer

11

Yusuke Kobayashi, Hiroyuki Nomura, Fumio Kataoka, Eiichiro Tominaga, Kouji Banno, and Daisuke Aoki

Abstract

Ovarian cancer has high morbidity and mortality worldwide, and its mortality is the highest among female genital cancers. Improvement in clinical outcome is an urgent issue for ovarian cancer, and clinical management including treatment has been repeatedly studied. Tumor markers such as CA125, CA19–9, and HE4 and image diagnosis using MRI, CT, PET, etc. are indispensable diagnostic aid tools. In the initial treatment, primary debulking surgery (PDS) is basically effective, but fertility preservation can be considered depending on the case. For advanced cancers, interval debulking surgery (IDS) is considered after several cycles of neoadjuvant chemotherapy (NAC). Although the combination of paclitaxel and carboplatin (TC) regimen was originally established as the standard chemotherapy for ovarian cancer regardless of tissue type, new treatments in clinical trials, including molecular-targeted therapeutic agents, are being investigated. The main treatment for recurrent cancer is chemotherapy, and sensitivity to chemotherapy based on the disease-free interval has been considered in the choice of treatment. Furthermore, in recent years, risk-reducing bilateral salpingo-oophorectomy has been undertaken to prevent the onset of hereditary breast and ovarian cancer. It is also being studied how to manage the follow-up after treatment and hormone replacement.

Keywords

Tumor marker · Diagnostic imaging · Staging laparotomy · Primary debulking surgery · Interval debulking surgery · Risk-reducing salpingo-oophorectomy · Adjuvant chemotherapy · Neoadjuvant chemotherapy · Molecular-targeted drugs · Follow-up after treatment · Hormone replacement therapy

Y. Kobayashi (✉) · H. Nomura · F. Kataoka · E. Tominaga · K. Banno · D. Aoki
Department of Obstetrics and Gynecology, Keio University School of Medicine, Tokyo, Japan
e-mail: kobax@a2.keio.jp

H. Katabuchi et al. (eds.), *Cell Biology of the Ovary*,
https://doi.org/10.1007/978-981-10-7941-2_11

11.1 Introduction

Ovarian cancer has high morbidity and mortality worldwide, and its mortality is the highest among female genital cancers. Improvement in clinical outcome is an urgent issue for ovarian cancer, and clinical management including treatment has been repeatedly studied. In this chapter, we introduce the latest clinical management of epithelial ovarian cancer with abundant literature.

11.1.1 Diagnosis

Ovarian cancer rarely causes subjective symptoms in the early stage and is difficult to perform preoperative histopathological diagnosis due to its anatomical location; therefore, tumor markers and imaging are used to diagnose ovarian cancer.

11.1.1.1 Tumor Markers

Tumor markers are commonly used when ovarian cancer is suspected before treatment and as a biomarker for responsiveness to chemotherapy and recurrence assessment. Cancer antigen 125 (CA125) is frequently determined, particularly for women with a risk of ovarian cancer. Menstruation, pregnancy, and endometriosis can cause false-positive results, but CA125 is used as a serum biomarker in many screening tests for ovarian cancer [1–4]. CA125 is a glycoprotein recognized by monoclonal antibody OC125, which is produced against the human serous adenocarcinoma-derived cell line OVCA433 [5]. In 2001, CA125 was identified as MUC16, a member of the mucin family of glycoproteins [6]. MUC16 binds to mesothelin, which is expressed in peritoneal mesothelioma and ovarian cancer cells, facilitates peritoneal metastasis [7], and inhibits the function of NK cells and antitumor immunity [8]. The positive rate of CA125 reaches 90% in FIGO Stage II–IV ovarian cancer but is about 50% in Stage I [9]. This rate is also higher in serous ovarian cancer than in other tissue types [10, 11]. CA125 is an indicator for therapeutic effect, and the half-life of CA125 is a prognostic factor for response to chemotherapy [12, 13]. Furthermore, the CA125 nadir value in patients who achieve a complete response in imaging findings after the first standard treatment is a useful index for recurrence, which correlates with progression-free survival (PFS) and overall survival (OS) [14, 15]. Since CA125 increases about 3–5 months before the condition where recurrence is detected in imaging diagnosis such as CT or MRI, CA125 seems to be useful as an indicator of recurrence [16, 17]. In contrast, early treatment of recurrence detected using CA125 does not contribute to outcomes, and there is a need for a thorough explanation of the interpretation of CA125 to patients [18]. Carbohydrate antigen 19–9 (CA19–9) is recognized by monoclonal antibody NS19–9, which was produced from the human colon cancer cell line SW1116 as an immunogen [19]. CA19–9 is likely to increase in borderline ovarian tumor and ovarian cancer in comparison with benign ovarian tumor; however, there is no relationship between CA19–9 levels and tissue types [20]. CA19–9 is found in CA125-negative patients with borderline ovarian tumor and ovarian cancer; therefore, CA19–9 is useful to distinguish between benign and malignant disease in these patients [20]. Human epididymis protein 4 (HE4) is a new marker that is encoded by the WAP four-disulfide core domain 2 (WFDC2) gene and has been shown to be overexpressed in ovarian cancer, and particularly in serous and endometrioid carcinoma, by comprehensive analyses of cDNA [21, 22]. In contrast to CA125, HE4 is not overexpressed in normal ovarian tissues, ovarian tumor, or low-grade tumor [23], which results in specificity for detection of ovarian cancer of 93%, which is higher than the 73% specificity of CA125 [24]. In addition, HE4 is detected in serum and urine of patients with ovarian cancer [25] and will be an important tumor marker for ovarian cancer based on accumulation of data in clinical practice.

11.1.1.2 Diagnostic Imaging

Diagnostic imaging using CT, MRI, and PET is useful for preoperative diagnosis and postoperative follow-up of patients with ovarian cancer. MRI is useful for local diagnosis before surgery, whereas CT is commonly used to assess metastasis and recurrence via rapid and widespread imaging. Also PET is particularly useful for diagnosis of recurrence. Ultrasound examination is the first option for diagnostic imaging of an adnexal mass and has diagnostic accuracy of 80–83% [26, 27]; however, diagnosis by ultrasonography is difficult in many patients. MRI shows good soft tissue contrast resolution and diagnostic accuracy of 89–93%, which is higher than that of ultrasonography [28, 29]. The diagnostic accuracy of CT is also relatively high, at 70–90% [30]; however, MRI with good contrast and without radiation exposure is generally conducted for local evaluation, and CT is used to determine the stage of ovarian cancer. PET is not recommended for detection and differential diagnosis of ovarian cancer because physiological accumulation is found in the normal ovary and accumulation is also detected in a benign tumor [31]. Based on these results, MRI is required for the most accurate diagnosis of patients with suspected ovarian cancer or in whom it is difficult to determine whether a tumor is benign or malignant by ultrasonography. Findings that are strong indicators of malignancy on MRI include necrosis in a solid component, cystic tumor in mural nodules, ascites, and peritoneal metastasis [28]. Contrast-enhanced MRI is useful to evaluate necrosis in solid and internal parts of ovarian tumors. Furthermore, in a dynamic study using chronological angiography via intravenous bolus injection of contrast agent, a malignant tumor is contrasted earlier due to greater blood flow to the tumor, compared to a tumor with low malignant potential. Consequently, contrast MRI is useful to evaluate solid tumor regions and to differentiate among benign, border, and malignant tumors [32]. Diffusion-weighted images (DWI) differ from T1-/T2-weighted images and are likely to show hyperintense signals for a malignant tumor with a high cell density and high *N*/*C* ratio [33]. Solid ovarian tumors also tend to have a reduced apparent diffusion coefficient (ADC) on DWI [34, 35]. It is difficult to differentiate between a benign and malignant ovarian tumor using DWI alone, but this imaging technique is useful as a contrast to T1-/T2-weighted images and for detection of disseminated lesions.

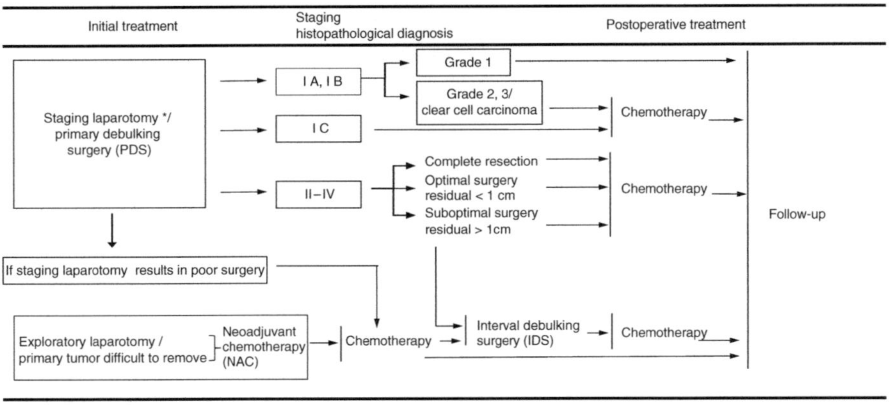

Fig. 11.1 Treatment of epithelial ovarian cancer (Modified from Komiyama S, et al. Int J Clin Oncol 2016; 21: 435–446). (*Asterisk*) Staging laparotomy—bilateral salpingo-oophorectomy + total hysterectomy + omentectomy + peritoneal cytology + pelvic/para-aortic lymph node dissection (biopsy) + biopsies from sites in the abdominal cavity

11.1.2 Surgical Therapy

Completion of surgery is a particularly important prognostic factor. Especially in advanced cancer, the residual tumor diameter after surgery correlates with outcomes, and many studies have shown a good prognosis for patients with a small residual tumor diameter who underwent complete surgery. Therefore, the principle of surgery is maximum tumor resection aimed at total extirpation of lesions. This procedure is referred to as primary debulking surgery (PDS). The objective of surgery includes determination of tissue type and staging (Fig. 11.1) [36].

11.1.2.1 Staging Laparotomy

For early ovarian cancer that is found to be limited to the ovary before and during operation, not only the affected side, contralateral adnexal resection and total hysterectomy (usually simple total hysterectomy) are recommended. In addition to peritoneal cytology (ascites or washing), omentectomy and peritoneal biopsy at various places in the peritoneal cavity are recommended to search for peritoneal dissemination. Lymphadenectomy and biopsy from the pelvic and para-aortic regions are also recommended based on the concern of retroperitoneal lymph node metastasis. However, staging laparotomy, which searches for the spread of cancers, is a technique recommended from the perspective of histopathologically determining the stage of progress and extracting cases in which postoperative treatment can be omitted. There is no evidence yet as to whether staging laparotomy itself directly improves prognosis.

11.1.2.2 Primary Debulking Surgery

The standard surgery for advanced cancer is primary debulking surgery (PDS), in which the goal is to extirpate intraperitoneal dissemination and metastatic lesions as much as possible. There is a correlation between residual tumor diameter and prognosis. Thus, PDS with maximum residual tumor diameters of <1 and ≥ 1 cm are defined as optimal and suboptimal surgeries, respectively, and an optimal surgery improves prognosis [37–39]. Recent studies have also indicated that a procedure resulting in the absence of a macroscopic residual tumor (complete surgery) significantly improves prognosis in comparison with an optimal surgery [40, 41]. However, few patients with advanced cancer can undergo the standard procedure, and the rate of optimal surgery is only 24–46% in Stage III patients [42]. No typical procedure for PDS has been established for patients with advanced cancer. The standard procedure is to resect tumors as much as possible regardless of dissemination or metastasis, in order to reduce tumors. Resection of peritoneal disseminated lesions, including the vesicouterine pouch, rectouterine pouch, Douglas' pouch, and paracolic gutter with the peripheral peritoneum, is recommended to reduce dissemination and metastatic lesions [43]. In case of invasion into the intestinal tract, intestinal segmentectomy and reconstruction should be proactively considered. In such cases, the patient should be informed of the possible need for colostomy, depending on the resection site [44, 45].

In patients with mucinous carcinoma, appendectomy is performed to differentiate from primary appendiceal carcinoma [46, 47]. If disseminated lesions in the diaphragm are found, stripping or full-thickness resection should be considered [48]. Resection of disseminated lesions in the diaphragm increases the rate of complete surgery [49]. If invasion of the spleen is found, splenectomy should be considered [50]. Even if disseminated lesions in the upper abdomen have progressed or expanded, patients in whom the tumor was macroscopically totally dissected have good prognosis; therefore, complete surgery should be considered proactively [40, 41]. Retroperitoneal lymph node dissection and biopsy have diagnostic significance in defining the correct advanced stage, but do not always have therapeutic significance. For patients with advanced ovarian cancer, retroperitoneal lymph node dissection should be considered if metastatic and disseminated lesions are surgically controlled.

11.1.2.3 Fertility Preservation Techniques

If a patient requests fertility preservation, basic procedures including adnexal resection on the affected side, omentectomy, and peritoneal cytology are essential. In patients with histological grade 1 ovarian cancer with no macroscopic invasion to the ovarian surface, capsule rupture, or peritoneal metastasis, microscopic metastasis to the opposite ovary is rare [51]. For prevention of infertility caused by decreased ovarian reserve and postoperative adhesion, it is permitted to omit biopsy in the opposite ovary if it is macroscopically normal. With regard to retroperitoneal lymph node dissection, the incidence of metastasis is low in patients with mucinous carcinoma and endometrioid carcinoma without pelvic progression or peritoneal metastasis [51, 52]. It is also permitted to perform biopsy alone in patients in whom fertility may be threatened by postoperative adhesion due to lymphadenectomy and who have less probability of metastasis. Procedures including staging laparotomy can be omitted by careful observation with gross appearance and palpation. Histopathological conditions to which fertility preservation is indicated are serous, mucinous, and endometrioid carcinoma, with Stage IA and grade 1 or 2. Those for which fertility preservation can be considered include non-particular type with Stage IC and grade 1 or 2 or clear cell carcinoma with Stage IA [53, 54].

11.1.2.4 Surgery in Elderly Patients

It is preferable for elderly patients to undergo debulking surgery with the aim of complete surgery. It is also important to plan the surgery considering systemic conditions, nutritional status, and complications. Careful attention should be paid to perioperative complications due to cardiopulmonary insufficiency in elderly patients [55]. Mortality within 30 days after operation for ovarian cancer is 1.5% in patients aged <70 years but increases to 6.6% and 9.8% in those aged 70–79 and ≥ 80 years, respectively. The causes of death include postoperative infection, hemorrhage, respiratory failure, heart failure, thrombus, and embolism [56]. Perioperative complications increase as surgical complications such as partial intestinal resection, diaphragmectomy, and splenectomy increase, so care must be taken for postoperative management [57]. Therefore, surgical procedures should be chosen based on systemic conditions, nutritional status, and disease stage at diagnosis, not based on age alone.

11.1.2.5 Risk-Reducing Salpingo-Oophorectomy

The lifetime incidence of ovarian cancer in women with BRCA1 and BRCA2 (BRCA1/BRCA2) mutations is high, and risk-reducing bilateral salpingo-oophorectomy (RRSO) is recommended to prevent hereditary breast and ovarian cancer (HBOC). RRSO decreases the incidence and mortality of ovarian cancer, fallopian tube cancer, and breast cancer in women with BRCA1/BRCA2 mutations [58, 59]. No definite conclusion on the timing of RRSO has been established; however, the 2017 National Comprehensive Cancer Network (NCCN) guidelines recommend RRSO at completion of delivery around 35–40 years old or the earliest age of diagnosis of ovarian cancer.

11.1.3 Chemotherapy

11.1.3.1 Neoadjuvant Chemotherapy

For patients with difficulty with total extirpation due to widespread peritoneal metastasis and metastatic lesions and those with critical complications including abundant ascites, poor systemic conditions, and thrombosis, interval debulking surgery (IDS) is an option after several cycles of neoadjuvant chemotherapy (NAC) (Fig. 11.1) [36]. Histological diagnosis may be inadequate due to chemotherapy modification by chemotherapy, and there was concern about the possibility that it cannot be diagnosed pathologically accurately with advanced ovarian cancer. Diagnostic laparoscopic surgery before NAC has predictive values of 100% for primary lesions and tissue types and 95% for disease stage [60]. Multiple randomized comparative studies of NAC + IDS and PDS have shown the value of NAC for advanced cancer [61–68]. Non-inferiority of NAC + IDS to PDS has been verified [61, 62], and NAC + IDS increases the optimal surgery rate [64, 65], reduces perioperative complications [61, 63, 65–67], and improves QOL [68] and OS [64]. Therefore, NAC + IDS is appropriate for cases in which complete surgery at first time is difficult and may be an option as standard therapy.

11.1.3.2 Adjuvant Chemotherapy

Cisplatin was approved as a key drug for ovarian cancer in 1980, and paclitaxel was launched in 1990. The combination of paclitaxel and cisplatin (TP) was confirmed as the standard regimen in the GOG111 and OV-10 studies. Since carboplatin is less toxic and more convenient to use compared to cisplatin, GOG158 and AGO trial was conducted and indicated that the efficacy of TC regimen was equal to that of TP regimen [69, 70]. The combination of paclitaxel and carboplatin (TC) regimen was originally established as the standard regimen for ovarian cancer in 2004 based on the results of the third International Ovarian Cancer Consensus Conference. A large-scale study GOG182-ICON5 using a new drug added to TC regimen was then conducted, but efficacy beyond that of TC regimen was not found [71], and TC regimen remained the first-line treatment for ovarian cancer. The administration method in TC regimen remains under consideration, and as a result of a randomized controlled trial JGOG 3016 comparing triweekly TC and dose-dense TC (dd-TC) to administer paclitaxel on a weekly basis, the progression of progression-free survival was significantly increased in the dd-TC group, so there is a possibility that dd-TC will be an international standard treatment in the future [72]. Regarding the route of administration, it is reported that intraperitoneal administration of cisplatin is more useful than intravenous administration [73–75]. However, intraperitoneal administration has not become standard treatment because there was no randomized comparison with current standard TC regimen. Since there is a report that intraperitoneal administration of carboplatin contributes to survival compared to intravenous administration [76–78], randomized phase II/III trials GOTIC/JGOG 3019 comparing only carboplatin administration route with dd-TC regimen are currently underway.

For patients with a suspected complication of peripheral neuropathy and those with alcohol intolerance, docetaxel-carboplatin (DC) regimen is an option because this regimen has no significant difference in response rate and PFS compared to TC regimen [79]. In patients with difficulty with administration of taxanes, carboplatin/pegylated liposomal doxorubicin (PLD-C) is also an option, based on the similar response rate, PFS, and OS compared to TC regimen [80]. In contrast, the response rates of clear cell and mucinous carcinoma are markedly lower than those of serous and endometrioid carcinoma; therefore, chemotherapy focusing on tissue types should be personalized. Clear cell carcinoma accounts for 24% of cases of ovarian cancer in Japan and is ranked second after serous carcinoma. However, in trials in Europe and the United States, the rates of clear cell carcinoma cases were approximately 8% [81], and almost no chemotherapy considering tissue type was examined. While TC regimen showed the response rates of clear cell carcinoma with 22–56% [82–84], the combination of irinotecan and cisplatin (CPT-P) regimen has been actively implemented in Japan [85–87]. An international collaborative comparative trial GCIG/JGOG3017 study was conducted, mainly in Japan due to the high incidence, and TC regimen and CPT-P regimen were compared with clear cell carcinoma. However, no significant difference was found in the contribution to prognosis [88]; thus TC regimen is currently recommended for clear cell cancer. Mucinous carcinoma accounts for approximately 10% of ovarian cancer, and it is difficult to distinguish between primary and metastasis especially in advanced cancers. It is necessary to pay attention to the possibility of metastasis from gastrointestinal cancer such as colorectal cancer [89]. The response rate to chemotherapy in patients diagnosed with primary ovarian mucinous carcinoma is extremely low, ranging from 13 to 26% [90, 91]. TC regimen is currently recommended for all tissue types, but new therapies are under development. For omission of postoperative chemotherapy, the 2016 NCCN guidelines specify, based on staging laparotomy, grade 1 serous/endometrioid with Stage IA–IB cases are recommended to observe and grade 2 with that cases are chemotherapy or follow-up. The 2013 National Institute for Health and Care Excellence (NICE) guidelines in the United Kingdom do not recommend postoperative chemotherapy for Stage IA or IB with grade 1 or 2 cases if staging laparotomy was conducted. Thus, omission of postoperative chemotherapy is possible in Stage IA and IB grade 1 cases determined by full staging laparotomy (Fig. 11.1) [36]. Also, in the case of grade 2, the risk of recurrence is low in Stage IA or IB if the staging laparotomy was performed, and there are reports that a good prognosis is obtained without postoperative chemotherapy, so it is considered possible depending on the case.

11.1.3.3 Molecular-Targeted Drugs in Chemotherapy

Bevacizumab is a molecular-targeted antibody drug against vascular endothelial growth factor (VEGF) and expected to have an additional effect when combined with chemotherapy. In the GOG218 and ICON7 studies, inclusion of bevacizumab in initial chemotherapy resulted in prolonged PFS [92, 93]. There are no results showing prolonged OS, but prolonged PFS with bevacizumab has also been shown in recurrent ovarian cancer, as described below. Bevacizumab can be expected to have an additional effect on existing chemotherapy in any situation of ovarian cancer treatment.

11.1.3.4 Chemotherapy in Elderly Patients

In addition to evaluating functions such as malnutrition, cognitive impairment, and mental status peculiar to the elderly, it is necessary to consider social support situations that can be hospitalized safely; however, in general, if the patient is eligible for surgery or a clinical trial qualified person, standard chemotherapy can be administered even for the elderly [94, 95]. During chemotherapy, it is important to improve anemia and control renal impairment.

11.1.4 Treatment of Recurrent Ovarian Cancer

Epithelial ovarian cancer responds to initial therapy, but more than a half of patients have a relapse. The common interval to relapse is within 2 years after initial treatment, and advanced Stage III/IV cases recur within 2 years in 55% of cases and within 5 years in >70% [81]. Median survival after relapse is approximately 2 years.

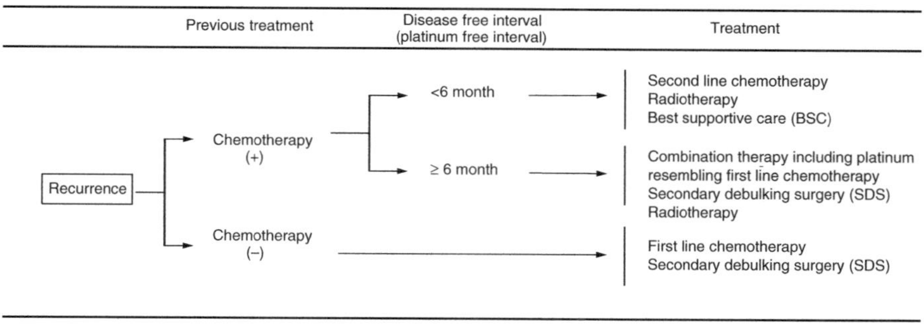

Fig. 11.2 Treatment of recurrent epithelial ovarian cancer (Modified from Komiyama S, et al. Int J Clin Oncol 2016; 21: 435–446)

Since complete cure is difficult for relapsed cases, the goal of treatment differs from that of initial treatment, with the goals of prolonged survival time, improvement in QOL, and symptom relief [96]. The main treatment for recurrent cancer is chemotherapy, and sensitivity to chemotherapy is important in the choice of treatment (Fig. 11.2) [36]. The interval from completion of initial chemotherapy to relapse (disease-free interval; DFI) correlates with the response rate to chemotherapy [97–100]. Thus, recurrent cancer with DFI ≥6 months is often categorized to platinum-sensitive and that with DFI <6 months is to platinum-resistant [101]. Taxanes are also evaluated in a similar manner [102].

For patients with drug-resistant cancer, it is necessary to select treatment without cross-resistance to initial treatment and choose a single agent considering toxicity [103, 104]. For patients with drug-sensitive cancer, multidrug combination therapy including platinum agent is selected [99, 100, 105]. However, chemotherapy has limitations, since the duration of response to chemotherapy for recurrent cancer does not exceed the duration of the initial chemotherapy [98]. Bevacizumab, an antibody against VEGF, is appropriate for recurrent ovarian cancer because it prolongs PFS in patients with platinum-sensitive (OCEANS study) and platinum-resistant (AURELIA study) recurrent ovarian cancer upon addition to current chemotherapy [106, 107]. In patients eligible for tumor resection, secondary debulking (cytoreductive) surgery may also be effective; however, this indication should be considered carefully [108]. Radiotherapy is mainly conducted to relieve symptoms [109, 110]. Recurrent cancer is refractory, and this limitation of treatment should be fully understood in the choice of procedures and indications. Proactive participation in clinical studies is necessary for development of new therapies.

11.1.5 Management After Treatment

11.1.5.1 Follow-Up After Treatment

In the 2016 NCCN guidelines, the follow-up interval after treatment is taken every 2–4 months for the first 2 years, every 3–6 months for the next 3 years, and every 1 year from the fifth year. The ESMO guidelines suggest a visit every 3 months for the first 2 years and every 6 months for fourth to fifth year [111]. Outcomes in a large-scale clinical trial showed median relapse-free survival is 22–29 months in high-risk cases of Stage I/II [112, 113] and 17–21 months in patients with advanced cancer [69, 70]; therefore, it is necessary to follow up every 3 months for the first 2 years. Recurrence occurred within 4 years in 95% of patients, and almost all had recurrence within 8 years [114]; therefore, after no recurrence for 5 years, the follow-up interval can be prolonged.

Regarding examination items to be performed as routines, the National Institutes of Health (NIH) consensus statement only specifies an interview, physical examination, vaginal and rectal examinations, and CA125 measurement. The ESMO clinical recommendations suggest interviews, physical examination, vaginal examination, and CA125 measurement every time, and CT scan is performed only when recurrence is suspected based on clinical findings or CA125 [115]. The 2016 NCCN guidelines recommend physical examination including pelvic examination and CA125 measurement as essential items every time and diagnostic imaging (CT, MRI, PET/CT) as indicated. In clinical practice, an interview, vaginal examination, and CA125 measurement are common, while diagnostic imaging is rare [116]. As described above, CA125 measurement is the main examination item; however, studies have shown that early treatment based on increased CA125 alone did not improve survival of patients with recurrent ovarian cancer [18] and that routine follow-up did not improve clinical outcomes based on the absence of a difference in survival between symptomatic and asymptomatic patients [117]. In contrast, another study showed that routine follow-up identified asymptomatic recurrent patients and improved the survival rate by surgery [118]. Thus, it remains unclear whether routine follow-up improves prognosis, and further accumulation of evidence is needed.

11.1.5.2 Hormone Replacement Therapy

The basic procedure for ovarian cancer includes bilateral oophorectomy, which may rapidly eliminate estrogen in premenopausal women; cause menopausal symptoms, dyslipidemia, and osteoporosis; and result in decreased QOL. >25% of patients with ovarian cancer are under 50 years old, in whom estrogen deficiency due to therapeutic oophorectomy is a concern. A study in non-cancer-bearing women showed that the survival rate in patients aged ≤45 years who underwent bilateral oophorectomy without estrogen replacement was lower than that in patients who did not undergo oophorectomy or underwent oophorectomy and estrogen replacement [119]. Correspondence to estrogen deficiency symptoms after oophorectomy is important for maintaining and improving QOL and may contribute to survival rate. HRT is one of the options for estrogen deficiency symptoms after oophorectomy. In the cohort study for the patients with invasive ovarian cancer and borderline ovarian tumors according to HRT use before and after diagnosis, the 5-year survival rate with HRT was significantly higher than that without HRT [120], whereas other studies showed no significant difference in recurrent, disease-free, or overall survival rates between patients with and without HRT [121–124]. Based on these results, HRT in patients with ovarian cancer is unlikely to increase the risk of recurrence and is recommended for patients in whom QOL is expected to be improved by estrogen replacement.

Conclusion

We outlined the latest clinical management of epithelial ovarian cancer. In surgical therapy, intestinal resection, resection of the peritoneum and diaphragm, and splenectomy are being actively performed in addition to the standard surgery. Furthermore, PARP (poly(ADP-ribose) polymerase) inhibitor against BRCA gene mutation-positive cases will be introduced as a molecular target therapy in the future, and it seems that it will become standard therapy. The gynecologic oncologist is not only required to be proficient in standard clinical management as described in this chapter but also to be familiar with new diagnosis and treatment methods.

References

1. Menon U, Gentry-Maharaj A, Hallett R, et al. Sensitivity and specificity of multimodal and ultrasound screening for ovarian cancer, and stage distribution of detected cancers: results of the prevalence screen of the UK Collaborative Trial of Ovarian Cancer Screening (UKCTOCS). Lancet Oncol. 2009;10:327–40.
2. Cramer DW, Bast RC Jr, Berg CD, et al. Ovarian cancer biomarker performance in prostate, lung, colorectal, and ovarian cancer screening trial specimens. Cancer Prev Res (Philadelphia, PA). 2011;4:365–74.
3. Skates SJ. Ovarian cancer screening: development of the risk of ovarian cancer algorithm (ROCA) and ROCA screening trials. Int J Gynecol Cancer. 2012;22:S24–6.
4. Drescher CW, Shah C, Thorpe J, et al. Longitudinal screening algorithm that incorporates change over time in CA125 levels identifies ovarian cancer earlier than a single-threshold rule. J Clin Oncol Off J Am Soc Clin Oncol. 2013;31:387–92.
5. Bast RC Jr, Feeney M, Lazarus H, Nadler LM, Colvin RB, Knapp RC. Reactivity of a monoclonal antibody with human ovarian carcinoma. J Clin Invest. 1981;68:1331–7.
6. Yin BW, Lloyd KO. Molecular cloning of the CA125 ovarian cancer antigen: identification as a new mucin, MUC16. J Biol Chem. 2001;276:27371–5.
7. Gubbels JA, Belisle J, Onda M, et al. Mesothelin-MUC16 binding is a high affinity. N-glycan dependent interaction that facilitates peritoneal metastasis of ovarian tumors. Mol Cancer. 2006;5:50.
8. Gubbels JA, Felder M, Horibata S, et al. MUC16 provides immune protection by inhibiting synapse formation between NK and ovarian tumor cells. Mol Cancer. 2010;9:11.
9. Gupta D, Lis CG. Role of CA125 in predicting ovarian cancer survival – a review of the epidemiological literature. J Ovarian Res. 2009;2:13.
10. Morales-Vasquez F, Pedernera E, Reynaga-Obregon J, et al. High levels of pretreatment CA125 are associated to improved survival in high grade serous ovarian carcinoma. J Ovarian Res. 2016;9:41.
11. Hogdall EV, Christensen L, Kjaer SK, et al. CA125 expression pattern, prognosis and correlation with serum CA125 in ovarian tumor patients. From The Danish "MALOVA" Ovarian Cancer Study. Gynecol Oncol. 2007;104:508–15.
12. Riedinger JM, Wafflart J, Ricolleau G, et al. CA 125 half-life and CA 125 nadir during induction chemotherapy are independent predictors of epithelial ovarian cancer outcome: results of a French multicentric study. Ann Oncol. 2006;17:1234–8.
13. Lee CK, Friedlander M, Brown C, et al. Early decline in cancer antigen 125 as a surrogate for progression-free survival in recurrent ovarian cancer. J Natl Cancer Inst. 2011;103:1338–42.
14. Krivak TC, Tian C, Rose GS, Armstrong DK, Maxwell GL. A Gynecologic Oncology Group Study of serum CA-125 levels in patients with stage III optimally debulked ovarian cancer treated with intraperitoneal compared to intravenous chemotherapy: an analysis of patients enrolled in GOG 172. Gynecol Oncol. 2009;115:81–5.

15. van Altena AM, Kolwijck E, Spanjer MJ, Hendriks JC, Massuger LF, de Hullu JA. CA125 nadir concentration is an independent predictor of tumor recurrence in patients with ovarian cancer: a population-based study. Gynecol Oncol. 2010;119:265–9.
16. Geurts SM, de Vegt F, van Altena AM, et al. Considering early detection of relapsed ovarian cancer: a review of the literature. Int J Gynecol Cancer. 2011;21:837–45.
17. Gadducci A, Cosio S, Zola P, Landoni F, Maggino T, Sartori E. Surveillance procedures for patients treated for epithelial ovarian cancer: a review of the literature. Int J Gynecol Cancer. 2007;17:21–31.
18. Rustin GJ, van der Burg ME, Griffin CL, et al. Early versus delayed treatment of relapsed ovarian cancer (MRC OV05/EORTC 55955): a randomised trial. Lancet (London). 2010;376:1155–63.
19. Koprowski H, Steplewski Z, Mitchell K, Herlyn M, Herlyn D, Fuhrer P. Colorectal carcinoma antigens detected by hybridoma antibodies. Somatic Cell Genet. 1979;5:957–71.
20. Cho HY, Kyung MS. Serum CA19-9 as a predictor of malignancy in primary ovarian mucinous tumors: a matched case-control study. Med Sci Monitor. 2014;20:1334–9.
21. Clauss A, Lilja H, Lundwall A. A locus on human chromosome 20 contains several genes expressing protease inhibitor domains with homology to whey acidic protein. Biochem J. 2002;368:233–42.
22. Drapkin R, von Horsten HH, Lin Y, et al. Human epididymis protein 4 (HE4) is a secreted glycoprotein that is overexpressed by serous and endometrioid ovarian carcinomas. Cancer Res. 2005;65:2162–9.
23. Galgano MT, Hampton GM, Frierson HF Jr. Comprehensive analysis of HE4 expression in normal and malignant human tissues. Modern Pathol. 2006;19:847–53.
24. Ferraro S, Braga F, Lanzoni M, Boracchi P, Biganzoli EM, Panteghini M. Serum human epididymis protein 4 vs carbohydrate antigen 125 for ovarian cancer diagnosis: a systematic review. J Clin Pathol. 2013;66:273–81.
25. Jia MM, Deng J, Cheng XL, et al. Diagnostic accuracy of urine HE4 in patients with ovarian cancer: a meta-analysis. Oncotarget. 2017;8:9660–71.
26. Kinkel K, Hricak H, Lu Y, Tsuda K, Filly RA. US characterization of ovarian masses: a meta-analysis. Radiology. 2000;217:803–11.
27. Bazot M, Darai E, Nassar-Slaba J, Lafont C, Thomassin-Naggara I. Value of magnetic resonance imaging for the diagnosis of ovarian tumors: a review. J Comput Assist Tomogr. 2008;32:712–23.
28. Hricak H, Chen M, Coakley FV, et al. Complex adnexal masses: detection and characterization with MR imaging--multivariate analysis. Radiology. 2000;214:39–46.
29. Bazot M, Nassar-Slaba J, Thomassin-Naggara I, Cortez A, Uzan S, Darai E. MR imaging compared with intraoperative frozen-section examination for the diagnosis of adnexal tumors; correlation with final histology. Eur Radiol. 2006;16:2687–99.
30. Moyle P, Addley HC, Sala E. Radiological staging of ovarian carcinoma. Semin Ultrasound CT MR. 2010;31:388–98.
31. Iyer VR, Lee SI. MRI, CT, and PET/CT for ovarian cancer detection and adnexal lesion characterization. AJR Am J Roentgenol. 2010;194:311–21.
32. Thomassin-Naggara I, Balvay D, Aubert E, et al. Quantitative dynamic contrast-enhanced MR imaging analysis of complex adnexal masses: a preliminary study. Eur Radiol. 2012;22:738–45.
33. Koh DM, Collins DJ. Diffusion-weighted MRI in the body: applications and challenges in oncology. AJR Am J Roentgenol. 2007;188:1622–35.
34. Thomassin-Naggara I, Darai E, Cuenod CA, et al. Contribution of diffusion-weighted MR imaging for predicting benignity of complex adnexal masses. Eur Radiol. 2009;19:1544–52.
35. Takeuchi M, Matsuzaki K, Nishitani H. Diffusion-weighted magnetic resonance imaging of ovarian tumors: differentiation of benign and malignant solid components of ovarian masses. J Comput Assist Tomogr. 2010;34:173–6.
36. Komiyama S, Katabuchi H, Mikami M, et al. Japan Society of Gynecologic Oncology guidelines 2015 for the treatment of ovarian cancer including primary peritoneal cancer and fallopian tube cancer. Int J Clin Oncol. 2016;21:435–46.

37. Bristow RE, Tomacruz RS, Armstrong DK, Trimble EL, Montz FJ. Survival effect of maximal cytoreductive surgery for advanced ovarian carcinoma during the platinum era: a meta-analysis. J Clin Oncol Off J Am Soc Clin Oncol. 2002;20:1248–59.
38. Panici PB, Maggioni A, Hacker N, et al. Systematic aortic and pelvic lymphadenectomy versus resection of bulky nodes only in optimally debulked advanced ovarian cancer: a randomized clinical trial. J Natl Cancer Inst. 2005;97:560–6.
39. Crawford SC, Vasey PA, Paul J, Hay A, Davis JA, Kaye SB. Does aggressive surgery only benefit patients with less advanced ovarian cancer? Results from an international comparison within the SCOTROC-1 Trial. J Clin Oncol Off J Am Soc Clin Oncol. 2005;23:8802–11.
40. Chi DS, Eisenhauer EL, Zivanovic O, et al. Improved progression-free and overall survival in advanced ovarian cancer as a result of a change in surgical paradigm. Gynecol Oncol. 2009;114:26–31.
41. du Bois A, Reuss A, Pujade-Lauraine E, Harter P, Ray-Coquard I, Pfisterer J. Role of surgical outcome as prognostic factor in advanced epithelial ovarian cancer: a combined exploratory analysis of 3 prospectively randomized phase 3 multicenter trials: by the Arbeitsgemeinschaft Gynaekologische Onkologie Studiengruppe Ovarialkarzinom (AGO-OVAR) and the Groupe d'Investigateurs Nationaux Pour les Etudes des Cancers de l'Ovaire (GINECO). Cancer. 2009;115:1234–44.
42. Chi DS, Eisenhauer EL, Lang J, et al. What is the optimal goal of primary cytoreductive surgery for bulky stage IIIC epithelial ovarian carcinoma (EOC)? Gynecol Oncol. 2006;103:559–64.
43. Eisenhauer EL, Abu-Rustum NR, Sonoda Y, et al. The addition of extensive upper abdominal surgery to achieve optimal cytoreduction improves survival in patients with stages IIIC-IV epithelial ovarian cancer. Gynecol Oncol. 2006;103:1083–90.
44. Bristow RE, Peiretti M, Zanagnolo V, Salani R, Giuntoli RL II, Maggioni A. Transverse colectomy in ovarian cancer surgical cytoreduction: operative technique and clinical outcome. Gynecol Oncol. 2008;109:364–9.
45. Juretzka MM, Barakat RR. Pelvic cytoreduction with rectosigmoid resection. Gynecol Oncol. 2007;104:40–4.
46. Ayhan A, Gultekin M, Taskiran C, et al. Routine appendectomy in epithelial ovarian carcinoma: is it necessary? Obstet Gynecol. 2005;105:719–24.
47. Dietrich CS III, Desimone CP, Modesitt SC, et al. Primary appendiceal cancer: gynecologic manifestations and treatment options. Gynecol Oncol. 2007;104:602–6.
48. Dowdy SC, Loewen RT, Aletti G, Feitoza SS, Cliby W. Assessment of outcomes and morbidity following diaphragmatic peritonectomy for women with ovarian carcinoma. Gynecol Oncol. 2008;109:303–7.
49. Fanfani F, Fagotti A, Gallotta V, et al. Upper abdominal surgery in advanced and recurrent ovarian cancer: role of diaphragmatic surgery. Gynecol Oncol. 2010;116:497–501.
50. Eisenkop SM, Spirtos NM, Lin WC. Splenectomy in the context of primary cytoreductive operations for advanced epithelial ovarian cancer. Gynecol Oncol. 2006;100:344–8.
51. Marpeau O, Schilder J, Zafrani Y, et al. Prognosis of patients who relapse after fertility-sparing surgery in epithelial ovarian cancer. Ann Surg Oncol. 2008;15:478–83.
52. Cho YH, Kim DY, Kim JH, et al. Is complete surgical staging necessary in patients with stage I mucinous epithelial ovarian tumors? Gynecol Oncol. 2006;103:878–82.
53. Satoh T, Tsuda H, Kanato K, et al. A non-randomized confirmatory study regarding selection of fertility-sparing surgery for patients with epithelial ovarian cancer: Japan Clinical Oncology Group Study (JCOG1203). Jpn J Clin Oncol. 2015;45:595–9.
54. Satoh T, Hatae M, Watanabe Y, et al. Outcomes of fertility-sparing surgery for stage I epithelial ovarian cancer: a proposal for patient selection. J Clin Oncol Off J Am Soc Clin Oncol. 2010;28:1727–32.
55. Langstraat C, Aletti GD, Cliby WA. Morbidity, mortality and overall survival in elderly women undergoing primary surgical debulking for ovarian cancer: a delicate balance requiring individualization. Gynecol Oncol. 2011;123:187–91.
56. Gerestein CG, Damhuis RA, de Vries M, Reedijk A, Burger CW, Kooi GS. Causes of postoperative mortality after surgery for ovarian cancer. Eur J Cancer (Oxford, England: 1990). 2009;45:2799–803.

57. Lim MC, Kang S, Song YJ, Park SH, Park SY. Feasibility and safety of extensive upper abdominal surgery in elderly patients with advanced epithelial ovarian cancer. J Korean Med Sci. 2010;25:1034–40.
58. Rebbeck TR, Kauff ND, Domchek SM. Meta-analysis of risk reduction estimates associated with risk-reducing salpingo-oophorectomy in BRCA1 or BRCA2 mutation carriers. J Natl Cancer Inst. 2009;101:80–7.
59. Domchek SM, Friebel TM, Singer CF, et al. Association of risk-reducing surgery in BRCA1 or BRCA2 mutation carriers with cancer risk and mortality. JAMA. 2010;304:967–75.
60. Onda T, Kobayashi H, Nakanishi T, et al. Feasibility study of neoadjuvant chemotherapy followed by interval debulking surgery for stage III/IV ovarian, tubal, and peritoneal cancers: Japan Clinical Oncology Group Study JCOG0206. Gynecol Oncol. 2009;113:57–62.
61. Kehoe S, Hook J, Nankivell M, et al. Primary chemotherapy versus primary surgery for newly diagnosed advanced ovarian cancer (CHORUS): an open-label, randomised, controlled, non-inferiority trial. Lancet (London). 2015;386:249–57.
62. Vergote I, Trope CG, Amant F, et al. Neoadjuvant chemotherapy or primary surgery in stage IIIC or IV ovarian cancer. N Engl J Med. 2010;363:943–53.
63. Onda T, Satoh T, Saito T, et al. Comparison of treatment invasiveness between upfront debulking surgery versus interval debulking surgery following neoadjuvant chemotherapy for stage III/IV ovarian, tubal, and peritoneal cancers in a phase III randomised trial: Japan Clinical Oncology Group Study JCOG0602. Eur J Cancer (Oxford, England: 1990). 2016;64:22–31.
64. Kuhn W, Rutke S, Spathe K, et al. Neoadjuvant chemotherapy followed by tumor debulking prolongs survival for patients with poor prognosis in International Federation of Gynecology and Obstetrics Stage IIIC ovarian carcinoma. Cancer. 2001;92:2585–91.
65. Lee SJ, Kim BG, Lee JW, Park CS, Lee JH, Bae DS. Preliminary results of neoadjuvant chemotherapy with paclitaxel and cisplatin in patients with advanced epithelial ovarian cancer who are inadequate for optimum primary surgery. J Obstet Gynaecol Res. 2006;32:99–106.
66. Hegazy MA, Hegazi RA, Elshafei MA, et al. Neoadjuvant chemotherapy versus primary surgery in advanced ovarian carcinoma. World J Surg Oncol. 2005;3:57.
67. Giannopoulos T, Butler-Manuel S, Taylor A, Ngeh N, Thomas H. Clinical outcomes of neoadjuvant chemotherapy and primary debulking surgery in advanced ovarian carcinoma. Eur J Gynaecol Oncol. 2006;27:25–8.
68. Chan YM, Ng TY, Ngan HY, Wong LC. Quality of life in women treated with neoadjuvant chemotherapy for advanced ovarian cancer: a prospective longitudinal study. Gynecol Oncol. 2003;88:9–16.
69. Ozols RF, Bundy BN, Greer BE, et al. Phase III trial of carboplatin and paclitaxel compared with cisplatin and paclitaxel in patients with optimally resected stage III ovarian cancer: a Gynecologic Oncology Group study. J Clin Oncol Off J Am Soc Clin Oncol. 2003;21:3194–200.
70. du Bois A, Luck HJ, Meier W, et al. A randomized clinical trial of cisplatin/paclitaxel versus carboplatin/paclitaxel as first-line treatment of ovarian cancer. J Natl Cancer Inst. 2003;95:1320–9.
71. Bookman MA, Brady MF, McGuire WP, et al. Evaluation of new platinum-based treatment regimens in advanced-stage ovarian cancer: a Phase III Trial of the Gynecologic Cancer Intergroup. J Clin Oncol Off J Am Soc Clin Oncol. 2009;27:1419–25.
72. Katsumata N, Yasuda M, Takahashi F, et al. Dose-dense paclitaxel once a week in combination with carboplatin every 3 weeks for advanced ovarian cancer: a phase 3, open-label, randomised controlled trial. Lancet (London). 2009;374:1331–8.
73. Markman M, Bundy BN, Alberts DS, et al. Phase III trial of standard-dose intravenous cisplatin plus paclitaxel versus moderately high-dose carboplatin followed by intravenous paclitaxel and intraperitoneal cisplatin in small-volume stage III ovarian carcinoma: an intergroup study of the Gynecologic Oncology Group, Southwestern Oncology Group, and Eastern Cooperative Oncology Group. J Clin Oncol Off J Am Soc Clin Oncol. 2001;19:1001–7.
74. Armstrong DK, Bundy B, Wenzel L, et al. Intraperitoneal cisplatin and paclitaxel in ovarian cancer. N Engl J Med. 2006;354:34–43.

75. Jaaback K, Johnson N, Lawrie TA. Intraperitoneal chemotherapy for the initial management of primary epithelial ovarian cancer. Cochrane Database Syst Rev. 2016;2016:Cd005340.
76. Fujiwara K, Markman M, Morgan M, Coleman RL. Intraperitoneal carboplatin-based chemotherapy for epithelial ovarian cancer. Gynecol Oncol. 2005;97:10–5.
77. Fujiwara K, Sakuragi N, Suzuki S, et al. First-line intraperitoneal carboplatin-based chemotherapy for 165 patients with epithelial ovarian carcinoma: results of long-term follow-up. Gynecol Oncol. 2003;90:637–43.
78. Fujiwara K, Nagao S, Kigawa J, et al. Phase II study of intraperitoneal carboplatin with intravenous paclitaxel in patients with suboptimal residual epithelial ovarian or primary peritoneal cancer: a Sankai Gynecology Cancer Study Group Study. Int J Gynecol Cancer. 2009;19:834–7.
79. Vasey PA, Jayson GC, Gordon A, et al. Phase III randomized trial of docetaxel-carboplatin versus paclitaxel-carboplatin as first-line chemotherapy for ovarian carcinoma. J Natl Cancer Inst. 2004;96:1682–91.
80. Pignata S, Scambia G, Ferrandina G, et al. Carboplatin plus paclitaxel versus carboplatin plus pegylated liposomal doxorubicin as first-line treatment for patients with ovarian cancer: the MITO-2 randomized phase III trial. J Clin Oncol Off J Am Soc Clin Oncol. 2011;29:3628–35.
81. Heintz AP, Odicino F, Maisonneuve P, et al. Carcinoma of the ovary. FIGO 26th annual report on the results of treatment in gynecological cancer. Int J Gynaecol Obstetr. 2006;95:S161–92.
82. Ho CM, Huang YJ, Chen TC, et al. Pure-type clear cell carcinoma of the ovary as a distinct histological type and improved survival in patients treated with paclitaxel-platinum-based chemotherapy in pure-type advanced disease. Gynecol Oncol. 2004;94:197–203.
83. Takano M, Kikuchi Y, Yaegashi N, et al. Clear cell carcinoma of the ovary: a retrospective multicentre experience of 254 patients with complete surgical staging. Br J Cancer. 2006;94:1369–74.
84. Utsunomiya H, Akahira J, Tanno S, et al. Paclitaxel-platinum combination chemotherapy for advanced or recurrent ovarian clear cell adenocarcinoma: a multicenter trial. Int J Gynecol Cancer. 2006;16:52–6.
85. Takano M, Kikuchi Y, Yaegashi N, et al. Adjuvant chemotherapy with irinotecan hydrochloride and cisplatin for clear cell carcinoma of the ovary. Oncol Rep. 2006;16:1301–6.
86. Takano M, Sugiyama T, Yaegashi N, et al. Progression-free survival and overall survival of patients with clear cell carcinoma of the ovary treated with paclitaxel-carboplatin or irinotecan-cisplatin: retrospective analysis. Int J Clin Oncol. 2007;12:256–60.
87. Sugiyama T, Yakushiji M, Kamura T, et al. Irinotecan (CPT-11) and cisplatin as first-line chemotherapy for advanced ovarian cancer. Oncology. 2002;63:16–22.
88. Sugiyama T, Okamoto A, Enomoto T, et al. Randomized phase III trial of irinotecan plus cisplatin compared with paclitaxel plus carboplatin as first-line chemotherapy for ovarian clear cell carcinoma: JGOG3017/GCIG Trial. J Clin Oncol Off J Am Soc Clin Oncol. 2016;34:2881–7.
89. Seidman JD, Kurman RJ, Ronnett BM. Primary and metastatic mucinous adenocarcinomas in the ovaries: incidence in routine practice with a new approach to improve intraoperative diagnosis. Am J Surg Pathol. 2003;27:985–93.
90. Shimada M, Kigawa J, Ohishi Y, et al. Clinicopathological characteristics of mucinous adenocarcinoma of the ovary. Gynecol Oncol. 2009;113:331–4.
91. Hess V, A'Hern R, Nasiri N, et al. Mucinous epithelial ovarian cancer: a separate entity requiring specific treatment. J Clin Oncol Off J Am Soc Clin Oncol. 2004;22:1040–4.
92. Burger RA, Brady MF, Bookman MA, et al. Incorporation of bevacizumab in the primary treatment of ovarian cancer. N Engl J Med. 2011;365:2473–83.
93. Perren TJ, Swart AM, Pfisterer J, et al. A phase 3 trial of bevacizumab in ovarian cancer. N Engl J Med. 2011;365:2484–96.
94. Eisenhauer EL, Tew WP, Levine DA, et al. Response and outcomes in elderly patients with stages IIIC-IV ovarian cancer receiving platinum-taxane chemotherapy. Gynecol Oncol. 2007;106:381–7.
95. Hilpert F, du Bois A, Greimel ER, et al. Feasibility, toxicity and quality of life of first-line chemotherapy with platinum/paclitaxel in elderly patients aged >or =70 years with advanced ovarian cancer—a study by the AGO OVAR Germany. Ann Oncol. 2007;18:282–7.

96. Ozols RF. Systemic therapy for ovarian cancer: current status and new treatments. Semin Oncol. 2006;33:S3–11.
97. Colombo N, Gore M. Treatment of recurrent ovarian cancer relapsing 6–12 months post platinum-based chemotherapy. Crit Rev Oncol Hematol. 2007;64:129–38.
98. Markman M, Markman J, Webster K, et al. Duration of response to second-line, platinum-based chemotherapy for ovarian cancer: implications for patient management and clinical trial design. J Clin Oncol Off J Am Soc Clin Oncol. 2004;22:3120–5.
99. Parmar MK, Ledermann JA, Colombo N, et al. Paclitaxel plus platinum-based chemotherapy versus conventional platinum-based chemotherapy in women with relapsed ovarian cancer: the ICON4/AGO-OVAR-2.2 trial. Lancet (London). 2003;361:2099–106.
100. Pfisterer J, Vergote I, Du Bois A, Eisenhauer E. Combination therapy with gemcitabine and carboplatin in recurrent ovarian cancer. Int J Gynecol Cancer. 2005;15:36–41.
101. Harries M, Gore M. Part II: chemotherapy for epithelial ovarian cancer-treatment of recurrent disease. Lancet Oncol. 2002;3:537–45.
102. Gronlund B, Hogdall C, Hansen HH, Engelholm SA. Results of reinduction therapy with paclitaxel and carboplatin in recurrent epithelial ovarian cancer. Gynecol Oncol. 2001;83:128–34.
103. Buda A, Floriani I, Rossi R, et al. Randomised controlled trial comparing single agent paclitaxel vs epidoxorubicin plus paclitaxel in patients with advanced ovarian cancer in early progression after platinum-based chemotherapy: an Italian Collaborative Study from the Mario Negri Institute, Milan, G.O.N.O. (Gruppo Oncologico Nord Ovest) group and I.O.R. (Istituto Oncologico Romagnolo) group. Br J Cancer. 2004;90:2112–7.
104. Sehouli J, Stengel D, Oskay-Oezcelik G, et al. Nonplatinum topotecan combinations versus topotecan alone for recurrent ovarian cancer: results of a phase III study of the North-Eastern German Society of Gynecological Oncology Ovarian Cancer Study Group. J Clin Oncol Off J Am Soc Clin Oncol. 2008;26:3176–82.
105. Gonzalez-Martin AJ, Calvo E, Bover I, et al. Randomized phase II trial of carboplatin versus paclitaxel and carboplatin in platinum-sensitive recurrent advanced ovarian carcinoma: a GEICO (Grupo Espanol de Investigacion en Cancer de Ovario) study. Ann Oncol. 2005;16:749–55.
106. Aghajanian C, Blank SV, Goff BA, et al. OCEANS: a randomized, double-blind, placebo-controlled phase III trial of chemotherapy with or without bevacizumab in patients with platinum-sensitive recurrent epithelial ovarian, primary peritoneal, or fallopian tube cancer. J Clin Oncol Off J Am Soc Clin Oncol. 2012;30:2039–45.
107. Pujade-Lauraine E, Hilpert F, Weber B, et al. Bevacizumab combined with chemotherapy for platinum-resistant recurrent ovarian cancer: the AURELIA open-label randomized phase III trial. J Clin Oncol Off J Am Soc Clin Oncol. 2014;32:1302–8.
108. Bristow RE, Puri I, Chi DS. Cytoreductive surgery for recurrent ovarian cancer: a meta-analysis. Gynecol Oncol. 2009;112:265–74.
109. Tinger A, Waldron T, Peluso N, et al. Effective palliative radiation therapy in advanced and recurrent ovarian carcinoma. Int J Radiat Oncol Biol Phys. 2001;51:1256–63.
110. E C QM, Gallant V, Samant R. Effective palliative radiotherapy for symptomatic recurrent or residual ovarian cancer. Gynecol Oncol. 2006;102:204–9.
111. Ledermann JA, Raja FA, Fotopoulou C, Gonzalez-Martin A, Colombo N, Sessa C. Newly diagnosed and relapsed epithelial ovarian carcinoma: ESMO Clinical Practice Guidelines for diagnosis, treatment and follow-up. Ann Oncol. 2013;24:vi24–32.
112. Rubin SC, Wong GY, Curtin JP, Barakat RR, Hakes TB, Hoskins WJ. Platinum-based chemotherapy of high-risk stage I epithelial ovarian cancer following comprehensive surgical staging. Obstet Gynecol. 1993;82:143–7.
113. Gadducci A, Sartori E, Maggino T, et al. Analysis of failures after negative second-look in patients with advanced ovarian cancer: an Italian multicenter study. Gynecol Oncol. 1998;68:150–5.
114. Dembo AJ. Controversy over combination chemotherapy in advanced ovarian cancer: what we learn from reports of matured data. J Clin Oncol Off J Am Soc Clin Oncol. 1986;4:1573–6.

115. Aebi S, Castiglione M. Epithelial ovarian carcinoma: ESMO clinical recommendations for diagnosis, treatment and follow-up. Ann Oncol. 2008;19:ii14–6.
116. Harmandayan GZ, Gao F, Mutch DG, Virgo KS, Gibb RK, Johnson FE. Ovarian cancer patient surveillance after curative-intent initial treatment. Gynecol Oncol. 2011;120:205–8.
117. Gadducci A, Fuso L, Cosio S, et al. Are surveillance procedures of clinical benefit for patients treated for ovarian cancer?: A retrospective Italian multicentric study. Int J Gynecol Cancer. 2009;19:367–74.
118. Tanner EJ, Chi DS, Eisenhauer EL, Diaz-Montes TP, Santillan A, Bristow RE. Surveillance for the detection of recurrent ovarian cancer: survival impact or lead-time bias? Gynecol Oncol. 2010;117:336–40.
119. Rocca WA, Grossardt BR, de Andrade M, Malkasian GD, Melton LJ 3rd. Survival patterns after oophorectomy in premenopausal women: a population-based cohort study. Lancet Oncol. 2006;7:821–8.
120. Mascarenhas C, Lambe M, Bellocco R, et al. Use of hormone replacement therapy before and after ovarian cancer diagnosis and ovarian cancer survival. Int J Cancer. 2006;119:2907–15.
121. Guidozzi F, Daponte A. Estrogen replacement therapy for ovarian carcinoma survivors: a randomized controlled trial. Cancer. 1999;86:1013–8.
122. Eeles RA, Tan S, Wiltshaw E, et al. Hormone replacement therapy and survival after surgery for ovarian cancer. BMJ (Clin Res ed). 1991;302:259–62.
123. Ursic-Vrscaj M, Bebar S, Zakelj MP. Hormone replacement therapy after invasive ovarian serous cystadenocarcinoma treatment: the effect on survival. Menopause (New York, NY). 2001;8:70–5.
124. Li L, Pan Z, Gao K, et al. Impact of post-operative hormone replacement therapy on life quality and prognosis in patients with ovarian malignancy. Oncol Lett. 2012;3:244–9.

Molecular Targeted Therapy for Epithelial Ovarian Cancer

12

Hironori Tashiro and Hidetaka Katabuchi

Abstract

Epithelial ovarian cancer (EOC) is associated with a variety of molecular and pathologic characteristics and has a spectrum of biological behavior. The current management of EOC is developing from traditional chemotherapy using cytotoxic agents to modern therapy using molecularly targeted agents, as the result of advances in translational research. This evolution has been based on our increased understanding of molecular pathology and numerous clinical trials in the field of gynecologic oncology. Current and future promising targeted therapy approaches for EOC can be classified into two types: (1) specific molecular targeting, using an apoptosis inducer associated with the unique molecular characteristics of the tumor, and (2) more generalized targeting strategies using angiogenesis and immune checkpoint inhibitors, which interact with the tumor microenvironment. Representative agents used in these two approaches include, respectively, (1) poly(adenosine diphosphate-ribose) polymerase (PARP) inhibitors, such as olaparib and niraparib, and (2) the angiogenesis inhibitor bevacizumab and the immune checkpoint inhibitor nivolumab. In this review, we discuss the present status of these promising agents in regard to the molecular targeting of EOC.

Keywords

Epithelial ovarian carcinoma · High-grade serous carcinoma · BRCA mutation · PARP inhibitors · Olaparib · Niraparib · Tumor microenvironment · Angiogenesis inhibitor · Bevacizumab · Immune checkpoint inhibitor · Nivolumab

---

H. Tashiro (✉)
Department of Mother-Child Nursing, Faculty of Life Sciences, Kumamoto University, Kumamoto, Japan
e-mail: htashiro@kumamoto-u.ac.jp

H. Katabuchi
Department of Obstetrics and Gynecology, Faculty of Life Sciences, Kumamoto University, Kumamoto, Japan

H. Katabuchi et al. (eds.), *Cell Biology of the Ovary*,
https://doi.org/10.1007/978-981-10-7941-2_12

Table 12.1 Current and future promising targeted molecules and antibodies

	Agent	Targeted molecule	Antibody
Tumor	PARP inhibitor	PARP1, 2	Olaparib Niraparib
Tumor microenvironment	VEGF inhibitor	VEGF-A	Bevacizumab
	Immune check point inhibitor	PD-1	Nivolumab

12.1 Introduction

Epithelial ovarian cancer (EOC) is one of the deadliest gynecological malignancies and the fifth leading cause of cancer-related mortality in women globally [1]. More than half of EOC patients are diagnosed at an advanced stage. The standard therapy for patients with advanced EOC generally involves surgical debulking and the combination of paclitaxel and carboplatin as a first-line therapy. Even though clinical remission can be achieved in such patients, over 70% in fact relapse, with the majority of recurrent patients being resistant to chemotherapy. The development of such chemoresistance is thus a significant restriction for the use of traditional cytotoxic agents in EOC.

Recent findings from molecular pathologic studies, as well as comprehensive genomic analyses using next-generation sequencing and gene expression profiling, have been revealing the biological diversity of EOCs [2]. EOCs can thus be classified into many subtypes, with distinctive molecular pathologic characteristics and distinct tumor microenvironments. This has important implications for the design of therapy, and through the accumulation of this knowledge, molecularly targeted therapies have become an integral part of precision medicine strategies that take into account the heterogeneity of EOC.

In this review, we summarize and discuss molecular pathologic interactions in EOC, with a focus on two promising types of molecular targeting strategy: (1) specific molecular targeting, using an apoptosis inducer associated with the unique molecular characteristics of the tumor, and (2) more generalized molecular targeting, using angiogenesis and immune checkpoint inhibitors, which interact with the tumor microenvironment (Table 12.1).

12.2 Specific Targets Associated with Tumor Molecular Pathologic Characteristics

12.2.1 Molecular Pathologic Characteristics of EOC

EOCs can be histologically classified into low-grade serous carcinoma (LGSC), high-grade serous carcinoma (HGSC), endometrioid carcinoma (EC), clear cell carcinoma (CCC), seromucinous carcinoma (SMC), mucinous carcinoma (MC), and malignant Brenner tumors (MBTs). These classifications are associated with various manifestations, clinical courses, resistances to chemotherapy, and prognoses and are associated with different underlying molecular alterations [3]. Conveniently, the different histologic types can be grossly divided into Type I and Type II tumors, which correspond to two distinct models of pathogenesis and molecular phenotype. Type I tumors, which develop from benign lesions to carcinomas, progress in an indolent course and are subdivided into three groups: LGSCs, endometriosis-related tumors (ECs, CCCs, and SMCs), and others (MCs and MBTs). On the other hand, Type II tumors are highly aggressive, progress rapidly, are typically diagnosed at an advanced stage, and are mostly composed of HGSCs. Recently, it has been considered that a major portion of Type II tumors develop from serous tubal intraepithelial carcinomas (STICs) in the fallopian tube and, as a result, disseminate as carcinomas that involve the ovary and extraovarian sites [4].

In terms of their molecular characteristics, Type I tumors are characterized by frequent mutations of the *KRAS*, *BRAF*, *ERBB2*, *CTNNB1*, *PTEN*, *PIK3CA*, and *ARID1A* genes. In particular, Type I tumors lack *TP53* mutations. Type II tumors are characterized by a high frequency of homologous recombination deficiency, *TP53* mutation, activation of the PI3K/AKT signaling pathway, and inactivation of Rb and Nf1, concomitant with tremendous genetic instability and intra-tumor heterogeneity [2, 3, 5–8]. Data published from The Cancer Genome Atlas (TCGA) has revealed that HGSCs of Type II can be further categorized into "differentiated," "proliferative," "mesenchymal," and "immunoreactive" subtypes, based on their molecular genetic profiles [2].

Mutations of the *TP53* gene, which has multifarious functions and operates as a gatekeeper gene in normal cells, are common in all carcinomas, and it is similarly a key gene for Type II EOC tumors [3]. Although many targeting therapies focusing on *TP53*, including gene therapy, have been attempted, effective therapies have not yet succeeded in clinical studies. Other molecules, such as *ERBB2* and members of the PI3K/AKT signaling pathway, have been also considered as targets. Several clinical trials using those molecular targets have been performed to verify safety and potential clinical benefit, but therapies targeted against common molecules in EOCs have not yet been adapted for clinical use.

BRCA1 [9] and *BRCA2* [10] (*BRCA1/BRCA2*) mutations, although not common, are specific for Type II EOCs, predominately HGSCs. Germline mutations of *BRCA1/BRCA2* are associated with hereditary breast and ovarian cancer (HBOC). To date, germline mutations in *BRCA1/BRCA2* have been found in less than 10% of HGSCs, while somatic mutations are associated with a further 3% of HGSC cases. Furthermore, the *BRCA1* promoter is hypermethylated and silenced in around 10% of HGSCs [2, 11–15]. Although genetic and epigenetic alterations of *BRCA1/BRCA2* are not so common in other EOC subtypes, recent molecular studies have revealed that PARP inhibitors have promise as molecularly targeted agents in *BRCA1-/BRCA2*-associated tumors [14, 15].

12.2.2 Specific Molecular Targets for BRCA1/BRCA2 Tumors: PARP Inhibitors

It has been reported that EOC patients carrying *BRCA1/BRCA2* mutations are more sensitive to platinum-based chemotherapy than EOC patients without *BRCA1/BRCA2* mutations [16–18]. The proteins encoded by the *BRCA1/BRCA2* genes

(Brca1/Brca2) maintain the integrity of the genome by regulating DNA damage response and repair. The major DNA repair pathways are direct repair, mismatch repair (MMR), base excision repair (BER), nucleotide excision repair (NER), and double-strand break (DSB) recombination repair [19]. Mammalian cells use several mechanisms [20] to repair DSBs, in particular nonhomologous end joining (NHEJ) and homologous recombination. Brca1/Brca2 form a complex with Rad51 that has an established role in homologous recombination during cell proliferation [21, 22]. When homologous recombination occurs in *BRCA1-/BRCA2*-mutated cells, DNA repair systems are not effective, resulting in genomic instability. However, several studies have also provided evidence of a "BRCAness" phenotype arising from defective homologous recombination that is linked to mechanisms other than *BRCA1/BRCA2* germline mutation [2, 23].

Poly(adenosine diphosphate-ribose) polymerases (PARPs), which constitute a large family of 18 proteins, are associated with DNA repair [24]. When PARP1 and PARP2 are activated by DNA damage, they facilitate pathways of DNA repair involving single-strand breaks (SSBs) and BER [25]. In vitro and in vivo studies have shown that *BRCA1-/BRCA2*-mutated cells are highly sensitive to PARP inhibition due to the loss of double repair systems [26] [27]. Basic research into PARP inhibition has been translated quickly into a novel tool for EOC treatment in the clinic. Recent trials investigating the efficacy of the PARP inhibitors, olaparib, talazoparib, niraparib, and veliparib, have shown clinical antitumor efficacy and tolerability [25]. As the first PARP inhibitor among them, the randomized, double-blind, phase II study of maintenance treatment with the oral PARP inhibitor olaparib (Study 19; NCT00753545) showed significant improvement in progression-free survival (PFS) in patients with recurrent, platinum-sensitive HGSCs carrying *BRCA1/BRCA2* mutations [28–30]. Two phase III trials of olaparib maintenance monotherapy for EOC patients with *BRCA1/BRCA2* mutations are ongoing: SOLO1 (NCT01844986) and SOLO2 (NCT01874353). In the SOLO2 study, which is no longer recruiting patients, the investigator-assessed median PFS was significantly longer with olaparib than with placebo. As a result, olaparib was granted accelerated approval by the US Food and Drug Administration (FDA) in December 2014, as a therapy for the treatment of patients with germline *BRCA1-/BRCA2*-mutated advanced EOC, especially HGSC, who have received three or more prior lines of chemotherapy [31]. This approval represents the first "personalized" therapy for EOC in the USA [25, 32].

Niraparib, which is also a PARP inhibitor, was approved by the FDA in March 2017. In the randomized, double-blind, phase III trial of niraparib maintenance therapy (NCT01847274), patients with platinum-sensitive, recurrent EOC were categorized according to the presence or absence of a germline *BRCA1* or *BRCA2* mutation (gBRCA and non-gBRCA cohorts, respectively), as well as the "BRCAness" of those in the non-gBRCA cohort (i.e., the presence or absence of a homologous recombination deficiency phenotype) [33]; patients were randomly assigned in a 2:1 ratio to receive niraparib or placebo once daily. Patients within the gBRCA cohort receiving niraparib had a significantly longer median PFS than those in the placebo group. Niraparib treatment also resulted in significantly longer PFS than placebo in both the homologous recombination deficiency positive subgroup of the non-gBRCA cohort and in the overall non-gBRCA cohort.

PARP inhibitors are relatively well tolerated for the treatment of EOC, either as single agents or in combination with chemotherapy, because they are oral agents. The main adverse effects (AEs) common to all PARP inhibitors are gastrointestinal symptoms, fatigue, and anemia. The frequent AEs of olaparib are nausea, fatigue, vomiting, taste alteration, anorexia, and anemia [34–37], while those of niraparib are thrombocytopenia, anemia, and neutropenia. These AEs can be effectively managed with dose modification [38].

12.3 Common Molecular Targets Related to the Tumor Microenvironment

12.3.1 Molecular Targets for the Tumor Microenvironment

The concept that the microenvironment of a developing tumor is a crucial regulator of carcinogenesis was originally proposed by Paget in his "seed and soil" hypothesis [39]. The tumor microenvironment is composed of blood vessels; immune cells (neutrophils, lymphocytes, macrophages, and dendritic cells); a complex tumor stroma that includes stromal cells, fibroblasts, and extracellular matrix; and a variety of signal network molecules that facilitate interactions between tumor and non-tumor cells. These interactions mediate the processes of angiogenesis, immune escape, and cancer stem cell and tumor progression, including dissemination [40, 41]. As such, the tumor microenvironment represents a complex but well-organized interplay between tumor cells, immune cells, and the vascular endothelium [42, 43] (Fig. 12.1). In regard to tumor cell dissemination in advanced EOC, ascites outside of the tumor mass also make a unique microenvironment, consisted of floating tumor cells and immune cells that produce various cytokines interacting with each other [44, 45]. The concept of the cancer stem cell itself is discussed in other chapters 6. In this section, we focus on the molecular targeting of angiogenesis and immune checkpoint pathways for EOC therapy.

12.3.1.1 Angiogenesis: Vascular Endothelial Growth Factor

Angiogenesis consists of a complex process regulated by a balance between pro- and anti-angiogenic factors in the microenvironment. Proangiogenic factors are vascular endothelial growth factor (VEGF), fibroblast growth factor (FGF), and platelet-derived growth factor (PDGF). VEGF is the most potent proangiogenic factor and is important for normal ovarian physiology and the reproductive cycle [46, 47]. VEGF is also a critical factor for tumor growth and survival, as it regulates the oxygen and nutrient supply necessary to maintain tumor cell function [48]. VEGF is expressed at high levels in EOC and at even higher levels in advanced carcinomas [49, 50]. VEGF induces ascites in patients with EOC, due to an increased peritoneal permeability [51], and experimental models have shown a positive correlation between VEGF expression and ascites [52, 53].

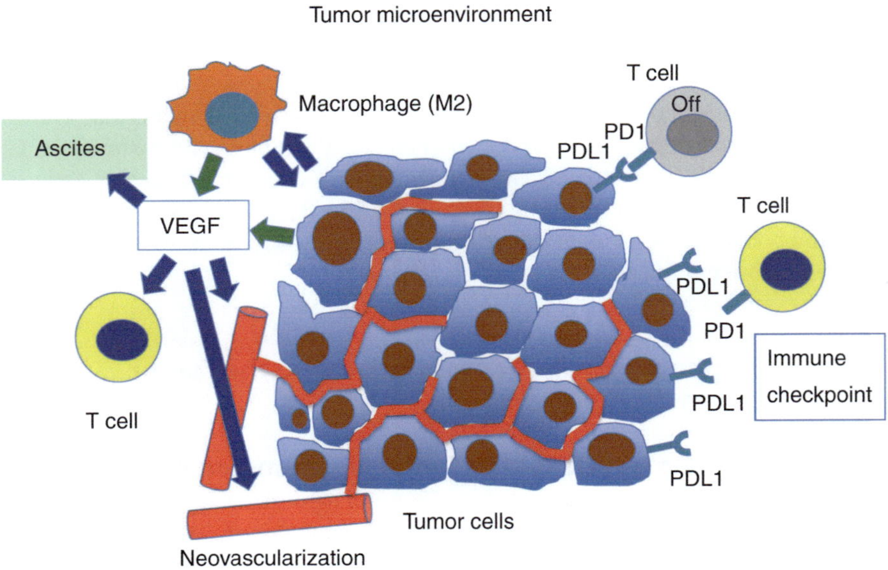

Fig. 12.1 Schema of the interrelationship between tumor and tumor microenvironment in epithelial ovarian cancer. *VEGF* vascular endothelial growth factor, *PD-1* programmed cell death 1, *PD-L1* programmed cell death ligand 1

At the molecular level, the VEGF receptor (VEGFR) activates several signaling pathways, including PI3K/AKT signaling, and promotes tumor growth, survival, and metastasis [54]. The VEGF family consists of seven ligands (VEGF-A to VEGF-E and the placental growth factors 1 and 2) and three receptors (VEGFR1–VEGFR3), which are cell surface tyrosine kinase receptors. VEGF-A, which is the primary regulator of angiogenesis, augments vascular permeability, mediates endothelial cell proliferation and migration, and allows cells to evade apoptosis. VEGF-A has multiple isoforms, of which, VEGF-A165 is commonly expressed in tumors [55]. VEGF-A, VEGF-C, and VEGF-D bind specifically to VEGFRs, of which VEGFR2 is the most important mediator of VEGF angiogenesis effects [56]. Direct inhibition of VEGF-A using monoclonal antibodies (e.g., bevacizumab), as well as small molecules inhibiting VEGFR (e.g., cediranib) and other receptor tyrosine kinases (e.g., trebananib), is a therapeutic strategy that has all been developed [57].

With its initial approval in the USA for advanced colorectal cancer in February 2004, the monoclonal VEGF antibody bevacizumab became the first anti-angiogenic therapy available for the treatment of patients with advanced cancer. To date, bevacizumab has been approved for the targeting of various types of advanced cancer, including breast cancer, colorectal cancer, non-small cell lung cancer, renal cancer, and malignant brain tumors.

Clinical trials using bevacizumab in combination with standard chemotherapy for EOC patients include the *International Collaborative Ovarian Neoplasm Group 7 (ICON7)* [58] and the *Gynecologic Oncology Group 218 (GOG218)* [59] studies for first-line therapy, and the *Ovarian Cancer Study Comparing Efficacy and Safety of Chemotherapy and Antiangiogenic Therapy in Platinum-Sensitive Recurrent Disease (OCEANS)* [60] and the *Avastin Use in Platinum-Resistant Epithelial Ovarian Cancer (AURELIA)* [61] studies. These have reported significant improvements in PFS but not in overall survival (OS) [62]. Furthermore, the improvement in a subgroup of patients with ascites, who have a poor prognosis, is noteworthy [63]. The Japanese Ministry of Health, Labour and Welfare approved bevacizumab for patients with EOC in November 2013. This approval was mainly based on the results of the GOG-0218 study conducted in the USA, Japan, and other countries, for first-line therapy of EOC. The GOG-0218 study investigated the efficacy and safety of bevacizumab when administered in combination with carboplatin/paclitaxel, a standard chemotherapy, and then bevacizumab monotherapy as a continuation therapy. In November 2014, the FDA approved bevacizumab in combination with chemotherapy for the treatment of patients with platinum-resistant recurrent EOC, based on the results from the AURELIA study, and expanded this approval in December 2016 for those with platinum-sensitive recurrent EOC, based on findings of the GOG-218 and OCEANS studies. In current clinical practice, therefore, bevacizumab is the most representative agent for molecularly targeted therapy in EOC. Furthermore, patients that were enrolled on the ICON7 trial were stratified into four TCGA molecular subtypes [2] to assess the association between molecular subtype and the efficacy of randomly assigned therapy with bevacizumab. The results demonstrated that in EOC, particularly in HGSC, molecular subtypes with the poorest survival (proliferative and mesenchymal subtypes) derive a comparably greater benefit from treatment that includes bevacizumab. The validation of these findings in an independent cohort would thus enable the use of bevacizumab in patient subgroups that are most likely to benefit [64].

The common AEs observed with VEGF-related inhibitors are: hypertension, proteinuria, impaired wound healing, gastrointestinal perforation, hypothyroidism, coagulation disorders (hemorrhage and thrombosis), reversible posterior leukoencephalopathy, cardiac impairment, and endocrine dysfunction. The role of the VEGF/VEGFR pathway in normal tissues has been reviewed to provide a molecular mechanism that links these diverse AEs with the physiological activity of VEGF/VEGFR [65, 66].

12.3.1.2 The Immune System: Immune Checkpoint Inhibitors

Remarkable advances have been made in our understanding of the interaction between immune cells and tumor cells in the tumor microenvironment during the last two decades [67–69]. Tumor cells contain nonsynonymous somatic mutations that alter the amino acid sequence of the proteins encoded by the affected genes [70]. Those alterations are foreign to the immune system and may therefore represent tumor-specific neoantigens capable of inducing antitumor immune responses [71, 72]. Activated T cells are the primary mediators of immune effector responses to tumor antigens; however tumor cells have the ability to escape the immune system. In recent years, most attention in regard to tackling this problem has been paid to the immune checkpoint, which represents an endogenous inhibitory signal for regulating the antigen recognition of T cell receptors during immune responses.

These checkpoints thus present the destruction of endogenous targets by tempering peripheral immune responses. As such, immune checkpoint-targeting therapy is an approach that can harness the functioning immune system in the tumor microenvironment. Two important suppressive immune checkpoints have been reported: (1) signaling via programmed cell death 1 (PD-1) and one of its ligands, PD-L1 (the PD-1/PD-L1 pathway) [73–75], and (2) signaling via CTLA-4 and its ligands B7–1 or B7–2 (the CTLA-4/B7 pathway) [76, 77]. PD-1/PD-L1 receptor-ligand signaling is a dominant immune checkpoint pathway in the tumor microenvironment that normally controls immune homeostasis but which can be induced by cancer cells to evade immune attack. Monoclonal antibodies that block this pathway have thus emerged as powerful weapons in the oncological armamentarium. Unlike PD-1/PD-L1 signaling, the CTLA-4 immune checkpoint predominantly operates early in the life cycle of the immune response, during T-cell priming and activation, and enhances the immunosuppressive activity of regulatory T cells.

Clinical trials of immune checkpoint inhibitors, such as anti-CTLA-4 antibodies, anti-PD-1 antibodies, and/or anti-PD-L1 antibodies, have shown efficacy against several kinds of malignant tumor, including melanomas and colorectal carcinomas. Whole-exome sequencing of melanomas treated with the anti-CTLA-4 antibodies ipilimumab or tremelimumab has revealed that these tumors have a very high mutational burden (0.5 to >100 mutations per megabase) compared with other malignant tumors [78]. In colorectal and non-colorectal carcinomas, it was shown that the effectiveness of the anti-PD-L1 antibody pembrolizumab is associated with the accumulation of mutations in genes of the DNA mismatch repair (MMR) pathway [79]. Whole-exome sequencing from this study revealed an average of 1782 somatic mutations per tumor in MMR-deficient carcinomas, as compared with only 73 mutations per tumor in patients with MMR-proficient carcinomas ($P = 0.007$) [79]. Accordingly, somatic mutation and neoantigen density has been shown to correlate with the long-term benefit of immune checkpoint blockade, suggesting that a high density of neoepitopes may enhance the clinical benefit from the blockade of immune checkpoints, which is designed to unleash endogenous responses to these mutation-associated neoantigens [78, 80].

A phase I trial of the anti-PD-L1 antibody BMS-936559 (NCT00729664) conducted in 207 patients with advanced solid tumors reported that out of the 17 EOC patients included in the cohort, 1/17 obtained partial response (PR), and 3/17 achieved stable disease (SD) [81]. More recent studies have looked at the efficacy of nivolumab as a fully humanized IgG4 that blocks the engagement of PD-1 by PD-1 ligands. A phase II trial of the efficacy and safety of single-agent nivolumab administered every 2 weeks at doses of 1 or 3 mg/kg in 20 patients with platinum-resistant EOC (two ten-patient cohorts tested in a 1:1 ratio) returned a highest overall response rate (RR) of 15% and a disease control rate (DCR) of 45%. In the 1 mg/kg cohort, one patient experienced a PR and four had SD. The objective RR (ORR) in this group was 10%, and the DCR was 50%. In the 3 mg/kg cohort, two patients had a complete response (CR) and two had SD. The ORR in this group was 20%, and the DCR was 40%. The median OS was prolonged to 20.0 months, while the median PFS for the two cohorts was 3.5 months. In a follow-up study, it was found that, after completing the initial 1 year of nivolumab treatment, two patients with CR survived without disease progression for >1 additional year in the absence of any adjuvant antitumor treatment [82–84]. EOC tissue from all 20 patients in this study was analyzed for tumor PD-L1 expression by immunohistochemistry. The results showed that 80% of tumor specimens had high expression of PD-L1; however the expression of PD-L1 did not significantly correlate with objective responses [82].

Immune checkpoints help to maintain tolerance to self-molecules, which is critical in preventing excessive autoimmunity. To this end, a number of immune checkpoint molecules exist that may serve to either augment or inhibit an immune response. The most common general AE for both anti-PD-1 and anti-PD-L1 antibodies is fatigue (observed in 21–71% of cases), which may present with early symptoms of hypothyroidism [85]. Other general AEs are pyrexia, chills, and infusion reactions. Organ-specific AEs are associated with dermatologic toxicities (maculopapular rash, Sweet's syndrome, follicular, and urticarial dermatitis), diarrhea/colitis, endocrine toxicities (hypophysitis, hypothyroidism, hyperthyroidism, thyroiditis, and adrenal insufficiency), hepatic toxicities (asymptomatic elevations in liver transaminases levels), and pneumonitis [85]. Ten percent of patients receiving immune-targeting therapy develop Grade 3 or 4 AEs, including dermatitis, pneumonia, colitis, hepatitis, vitiligo, and thyroiditis. Serious immune-related AEs are severe skin reactions (Stevens–Johnson syndrome and toxic epidermal necrolysis), acute interstitial pneumonia, acute respiratory distress syndrome, colitis with gastrointestinal perforation, anaphylactic shock, Type I diabetes, thrombocytopenia, neutropenia, encephalopathy and neurological sequelae, Guillain–Barré syndrome, myelitis and motor sequelae, myocarditis and cardiac insufficiency, acute adrenal insufficiency, and nephritis [83, 85, 86]. It has also been reported that nivolumab has a risk of embryofetal toxicity in the cynomolgus monkey, with females of reproductive age consequently advised to use effective contraception both during treatment with nivolumab and for at least 5 months after the last dose [83].

The durable RR in a subset of EOCs and the good tolerability of checkpoint inhibitors support further investigation; however, it is clear that only a proportion of patients will respond to checkpoint inhibitor monotherapy. In EOCs, data published from the TCGA have revealed that a subset of HGSCs have an "immunoreactive" subtype, based on their molecular genetic profiling [2, 87, 88]. This subtype is not only immunogenic but is also associated with *BRCA1* mutations, which lead to defects in homologous recombination DNA repair. Tumors with DNA repair deficiencies are thought to stimulate the immune system through the expression of neoantigens arising from their high mutational load, suggesting that these patients may be more responsive to immune checkpoint inhibitors.

An increased risk of EOC is also associated with Lynch syndrome [89]. In Lynch syndrome families with MMR mutations, EOCs developed at a mean age of 48 years and are associated with a higher incidence of endometrioid (35%) and clear cell (17%) histologic subtypes. EOCs with MMR defects may thus respond more favorably to immune checkpoint inhibitors. Indeed, future clinical trials of the therapies using immune checkpoint inhibitors and further exploration of biomarkers will be critical to advance our understanding of how best to select biomarker-driven therapies for EOC.

Conclusion

A little less than a dozen molecular alterations have been associated with a clinical benefit from targeted therapies, leading to the approval of targeted therapies for EOCs with diverse molecular features. Such tumor heterogeneity is thus a critical aspect of the design and optimization of personalized therapy approaches for EOC. A number of agents, including monoclonal antibodies, are advancing through clinical trials; however molecularly targeted therapies need to be improved to minimize AEs and prolong PFS and overall RR for EOC patients. Further study is required to improve our understanding of the biological significance of these approaches however, since the best clinical strategy for selecting effective targeted agents may involve more than one biomarker. Indeed, algorithms may be required to determine the appropriate biomarker combinations for each setting [83]. Moreover, beyond the already known molecular alterations, additional co-regulatory mechanisms and epigenetic pathways are unfolding as major players in tumorigenesis and offer themselves as targets for future consideration.

References

1. Ferlay J, Shin HR, Bray F, Forman D, Mathers C, Parkin DM. Estimates of worldwide burden of cancer in 2008: GLOBOCAN 2008. Int J Cancer. 2010;127:2893–917.
2. Cancer Genome Atlas Research Network. Integrated genomic analyses of ovarian carcinoma. Nature. 2011;474:609–15.
3. Tashiro H, Imamura Y, Motohara T, Sakaguchi I, Katabuchi H. Morphological and molecular pathogenesis of epithelial ovarian tumors. Frontiers in ovarian cancer science. Singapore: Springer Singapore; 2017.
4. Kurman RJ, Shih Ie M. The dualistic model of ovarian carcinogenesis: revisited, revised, and expanded. Am J Pathol. 2016;186:733–47.
5. Salani R, Kurman RJ, Giuntoli R II, Gardner G, Bristow R, Wang TL, et al. Assessment of TP53 mutation using purified tissue samples of ovarian serous carcinomas reveals a higher mutation rate than previously reported and does not correlate with drug resistance. Int J Gynecol Cancer. 2008;18:487–91.
6. Sasaki R, Narisawa-Saito M, Yugawa T, Fujita M, Tashiro H, Katabuchi H, et al. Oncogenic transformation of human ovarian surface epithelial cells with defined cellular oncogenes. Carcinogenesis. 2009;30:423–31.
7. Jazaeri AA, Bryant JL, Park H, Li H, Dahiya N, Stoler MH, et al. Molecular requirements for transformation of fallopian tube epithelial cells into serous carcinoma. Neoplasia. 2011;13:899–911.
8. Motohara T, Masuko S, Ishimoto T, Yae T, Onishi N, Muraguchi T, et al. Transient depletion of p53 followed by transduction of c-Myc and K-Ras converts ovarian stem-like cells into tumor-initiating cells. Carcinogenesis. 2011;32:1597–606.
9. Miki Y, Swensen J, Shattuck-Eidens D, Futreal PA, Harshman K, Tavtigian S, et al. A strong candidate for the breast and ovarian cancer susceptibility gene BRCA1. Science. 1994;266:66–71.
10. Wooster R, Bignell G, Lancaster J, Swift S, Seal S, Mangion J, et al. Identification of the breast cancer susceptibility gene BRCA2. Nature. 1995;378:789–92.
11. Cannistra SA. BRCA-1 in sporadic epithelial ovarian cancer: lessons learned from the genetics of hereditary disease. Clin Cancer Res. 2007;13:7225–7.
12. Wang C, Horiuchi A, Imai T, Ohira S, Itoh K, Nikaido T, et al. Expression of BRCA1 protein in benign, borderline, and malignant epithelial ovarian neoplasms and its relationship to methylation and allelic loss of the BRCA1 gene. J Pathol. 2004;202:215–23.

13. Baldwin RL, Nemeth E, Tran H, Shvartsman H, Cass I, Narod S, et al. BRCA1 promoter region hypermethylation in ovarian carcinoma: a population-based study. Cancer Res. 2000;60:5329–33.
14. Koczkowska M, Zuk M, Gorczynski A, Ratajska M, Lewandowska M, Biernat W, et al. Detection of somatic BRCA1/2 mutations in ovarian cancer - next-generation sequencing analysis of 100 cases. Cancer Med. 2016;5:1640–6.
15. Hennessy BT, Timms KM, Carey MS, Gutin A, Meyer LA, Flake DD II, et al. Somatic mutations in BRCA1 and BRCA2 could expand the number of patients that benefit from poly (ADP ribose) polymerase inhibitors in ovarian cancer. J Clin Oncol. 2010;28:3570–6.
16. Tan DS, Rothermundt C, Thomas K, Bancroft E, Eeles R, Shanley S, et al. "BRCAness" syndrome in ovarian cancer: a case-control study describing the clinical features and outcome of patients with epithelial ovarian cancer associated with BRCA1 and BRCA2 mutations. J Clin Oncol. 2008;26:5530–6.
17. Alsop K, Fereday S, Meldrum C, deFazio A, Emmanuel C, George J, et al. BRCA mutation frequency and patterns of treatment response in BRCA mutation-positive women with ovarian cancer: a report from the Australian Ovarian Cancer Study Group. J Clin Oncol. 2012;30:2654–63.
18. Vencken PM, Kriege M, Hoogwerf D, Beugelink S, van der Burg ME, Hooning MJ, et al. Chemosensitivity and outcome of BRCA1- and BRCA2-associated ovarian cancer patients after first-line chemotherapy compared with sporadic ovarian cancer patients. Ann Oncol. 2011;22:1346–52.
19. Plummer R. Perspective on the pipeline of drugs being developed with modulation of DNA damage as a target. Clin Cancer Res. 2010;16:4527–31.
20. Karran P. DNA double strand break repair in mammalian cells. Curr Opin Genet Dev. 2000;10:144–50.
21. Chen CF, Chen PL, Zhong Q, Sharp ZD, Lee WH. Expression of BRC repeats in breast cancer cells disrupts the BRCA2-Rad51 complex and leads to radiation hypersensitivity and loss of G(2)/M checkpoint control. J Biol Chem. 1999;274:32931–5.
22. Scully R, Chen J, Plug A, Xiao Y, Weaver D, Feunteun J, et al. Association of BRCA1 with Rad51 in mitotic and meiotic cells. Cell. 1997;88:265–75.
23. Konstantinopoulos PA, Spentzos D, Karlan BY, Taniguchi T, Fountzilas E, Francoeur N, et al. Gene expression profile of BRCAness that correlates with responsiveness to chemotherapy and with outcome in patients with epithelial ovarian cancer. J Clin Oncol. 2010;28:3555–61.
24. Ame JC, Spenlehauer C, de Murcia G. The PARP superfamily. BioEssays. 2004;26:882–93.
25. Dziadkowiec KN, Gasiorowska E, Nowak-Markwitz E, Jankowska A. PARP inhibitors: review of mechanisms of action and BRCA1/2 mutation targeting. Prz Menopauzalny. 2016;15:215–9.
26. Farmer H, McCabe N, Lord CJ, Tutt AN, Johnson DA, Richardson TB, et al. Targeting the DNA repair defect in BRCA mutant cells as a therapeutic strategy. Nature. 2005;434:917–21.
27. Bryant HE, Schultz N, Thomas HD, Parker KM, Flower D, Lopez E, et al. Specific killing of BRCA2-deficient tumours with inhibitors of poly(ADP-ribose) polymerase. Nature. 2005;434:913–7.
28. Ledermann J, Harter P, Gourley C, Friedlander M, Vergote I, Rustin G, et al. Olaparib maintenance therapy in platinum-sensitive relapsed ovarian cancer. N Engl J Med. 2012;366:1382–92.
29. Ledermann J, Harter P, Gourley C, Friedlander M, Vergote I, Rustin G, et al. Olaparib maintenance therapy in patients with platinum-sensitive relapsed serous ovarian cancer: a preplanned retrospective analysis of outcomes by BRCA status in a randomised phase 2 trial. Lancet Oncol. 2014;15:852–61.
30. Matulonis UA, Harter P, Gourley C, Friedlander M, Vergote I, Rustin G, et al. Olaparib maintenance therapy in patients with platinum-sensitive, relapsed serous ovarian cancer and a BRCA mutation: overall survival adjusted for postprogression poly(adenosine diphosphate ribose) polymerase inhibitor therapy. Cancer. 2016;122:1844–52.
31. Walsh CS. Two decades beyond BRCA1/2: homologous recombination, hereditary cancer risk and a target for ovarian cancer therapy. Gynecol Oncol. 2015;137:343–50.
32. Bai H, Cao D, Yang J, Li M, Zhang Z, Shen K. Genetic and epigenetic heterogeneity of epithelial ovarian cancer and the clinical implications for molecular targeted therapy. J Cell Mol Med. 2016;20:581–93.

33. Abiko K, Mandai M, Hamanishi J, Yoshioka Y, Matsumura N, Baba T, et al. PD-L1 on tumor cells is induced in ascites and promotes peritoneal dissemination of ovarian cancer through CTL dysfunction. Clin Cancer Res. 2013;19:1363–74.
34. Fong PC, Boss DS, Yap TA, Tutt A, Wu P, Mergui-Roelvink M, et al. Inhibition of poly(ADP-ribose) polymerase in tumors from BRCA mutation carriers. N Engl J Med. 2009;361:123–34.
35. Kaufman B, Shapira-Frommer R, Schmutzler RK, Audeh MW, Friedlander M, Balmana J, et al. Olaparib monotherapy in patients with advanced cancer and a germline BRCA1/2 mutation. J Clin Oncol. 2015;33:244–50.
36. Audeh MW, Carmichael J, Penson RT, Friedlander M, Powell B, Bell-McGuinn KM, et al. Oral poly(ADP-ribose) polymerase inhibitor olaparib in patients with BRCA1 or BRCA2 mutations and recurrent ovarian cancer: a proof-of-concept trial. Lancet. 2010;376:245–51.
37. Gelmon KA, Tischkowitz M, Mackay H, Swenerton K, Robidoux A, Tonkin K, et al. Olaparib in patients with recurrent high-grade serous or poorly differentiated ovarian carcinoma or triple-negative breast cancer: a phase 2, multicentre, open-label, non-randomised study. Lancet Oncol. 2011;12:852–61.
38. Mirza MR, Monk BJ, Herrstedt J, Oza AM, Mahner S, Redondo A, et al. Niraparib maintenance therapy in platinum-sensitive, recurrent ovarian cancer. N Engl J Med. 2016;375:2154–64.
39. Paget S. The distribution of secondary growths in cancer of the breast. Lancet. 1889;133:571–3.
40. Lokadasan R, James FV, Narayanan G, Prabhakaran PK. Targeted agents in epithelial ovarian cancer: review on emerging therapies and future developments. Ecancermedicalscience. 2016;10:626.
41. Albini A, Sporn MB. The tumour microenvironment as a target for chemoprevention. Nat Rev Cancer. 2007;7:139–47.
42. Carmeliet P, Jain RK. Molecular mechanisms and clinical applications of angiogenesis. Nature. 2011;473:298–307.
43. Pickup MW, Mouw JK, Weaver VM. The extracellular matrix modulates the hallmarks of cancer. EMBO Rep. 2014;15:1243–53.
44. Takaishi K, Komohara Y, Tashiro H, Ohtake H, Nakagawa T, Katabuchi H, et al. Involvement of M2-polarized macrophages in the ascites from advanced epithelial ovarian carcinoma in tumor progression via Stat3 activation. Cancer Sci. 2010;101:2128–36.
45. Kim S, Kim B, Song YS. Ascites modulates cancer cell behavior, contributing to tumor heterogeneity in ovarian cancer. Cancer Sci. 2016;107:1173–8.
46. Fraser HM, Wilson H, Morris KD, Swanston I, Wiegand SJ. Vascular endothelial growth factor Trap suppresses ovarian function at all stages of the luteal phase in the macaque. J Clin Endocrinol Metab. 2005;90:5811–8.
47. Fraser HM, Duncan WC. Vascular morphogenesis in the primate ovary. Angiogenesis. 2005;8:101–16.
48. Folkman J. Tumor angiogenesis: therapeutic implications. N Engl J Med. 1971;285:1182–6.
49. Abu-Jawdeh GM, Faix JD, Niloff J, Tognazzi K, Manseau E, Dvorak HF, et al. Strong expression of vascular permeability factor (vascular endothelial growth factor) and its receptors in ovarian borderline and malignant neoplasms. Lab Investig. 1996;74:1105–15.
50. Shen GH, Ghazizadeh M, Kawanami O, Shimizu H, Jin E, Araki T, et al. Prognostic significance of vascular endothelial growth factor expression in human ovarian carcinoma. Br J Cancer. 2000;83:196–203.
51. Herr D, Sallmann A, Bekes I, Konrad R, Holzheu I, Kreienberg R, et al. VEGF induces ascites in ovarian cancer patients via increasing peritoneal permeability by downregulation of Claudin 5. Gynecol Oncol. 2012;127:210–6.
52. Mesiano S, Ferrara N, Jaffe RB. Role of vascular endothelial growth factor in ovarian cancer: inhibition of ascites formation by immunoneutralization. Am J Pathol. 1998;153:1249–56.
53. Schumacher JJ, Dings RP, Cosin J, Subramanian IV, Auersperg N, Ramakrishnan S. Modulation of angiogenic phenotype alters tumorigenicity in rat ovarian epithelial cells. Cancer Res. 2007;67:3683–90.
54. Park ST, Kim BR, Park SH, Lee JH, Lee EJ, Lee SH, et al. Suppression of VEGF expression through interruption of the HIF1alpha and Akt signaling cascade modulates the antiangiogenic activity of DAPK in ovarian carcinoma cells. Oncol Rep. 2014;31:1021–9.

55. Niu G, Chen X. Vascular endothelial growth factor as an anti-angiogenic target for cancer therapy. Curr Drug Targets. 2010;11:1000–17.
56. Ferrara N, Gerber HP, LeCouter J. The biology of VEGF and its receptors. Nat Med. 2003;9:669–76.
57. Ferrara N, Kerbel RS. Angiogenesis as a therapeutic target. Nature. 2005;438:967–74.
58. Perren TJ, Swart AM, Pfisterer J, Ledermann JA, Pujade-Lauraine E, Kristensen G, et al. A phase 3 trial of bevacizumab in ovarian cancer. N Engl J Med. 2011;365:2484–96.
59. Burger RA, Brady MF, Bookman MA, Fleming GF, Monk BJ, Huang H, et al. Incorporation of bevacizumab in the primary treatment of ovarian cancer. N Engl J Med. 2011;365:2473–83.
60. Aghajanian C, Blank SV, Goff BA, Judson PL, Teneriello MG, Husain A, et al. OCEANS: a randomized, double-blind, placebo-controlled phase III trial of chemotherapy with or without bevacizumab in patients with platinum-sensitive recurrent epithelial ovarian, primary peritoneal, or fallopian tube cancer. J Clin Oncol. 2012;30:2039–45.
61. Pujade-Lauraine E, Hilpert F, Weber B, Reuss A, Poveda A, Kristensen G, et al. Bevacizumab combined with chemotherapy for platinum-resistant recurrent ovarian cancer: the AURELIA open-label randomized phase III trial. J Clin Oncol. 2014;32:1302–8.
62. Shaw D, Clamp A, Jayson GC. Angiogenesis as a target for the treatment of ovarian cancer. Curr Opin Oncol. 2013;25:558–65.
63. Pujade-Lauraine E, Ledermann JA, Selle F, Gebski V, Penson RT, Oza AM, et al. Olaparib tablets as maintenance therapy in patients with platinum-sensitive, relapsed ovarian cancer and a BRCA1/2 mutation (SOLO2/ENGOT-Ov21): a double-blind, randomised, placebo-controlled, phase 3 trial. Lancet Oncol. 2017;18:1274–84.
64. Kommoss S, Winterhoff B, Oberg AL, Konecny GE, Wang C, Riska SM, et al. Bevacizumab may differentially improve ovarian cancer outcome in patients with proliferative and mesenchymal molecular subtypes. Clin Cancer Res. 2017;23:3794–801.
65. Kamba T, McDonald DM. Mechanisms of adverse effects of anti-VEGF therapy for cancer. Br J Cancer. 2007;96:1788–95.
66. Roodhart JM, Langenberg MH, Witteveen E, Voest EE. The molecular basis of class side effects due to treatment with inhibitors of the VEGF/VEGFR pathway. Curr Clin Pharmacol. 2008;3:132–43.
67. Liyanage UK, Moore TT, Joo HG, Tanaka Y, Herrmann V, Doherty G, et al. Prevalence of regulatory T cells is increased in peripheral blood and tumor microenvironment of patients with pancreas or breast adenocarcinoma. J Immunol. 2002;169:2756–61.
68. Chambers CA, Kuhns MS, Egen JG, Allison JP. CTLA-4-mediated inhibition in regulation of T cell responses: mechanisms and manipulation in tumor immunotherapy. Annu Rev Immunol. 2001;19:565–94.
69. Linehan DC, Goedegebuure PS. CD25+ CD4+ regulatory T-cells in cancer. Immunol Res. 2005;32:155–68.
70. Vogelstein B, Papadopoulos N, Velculescu VE, Zhou S, Diaz LA Jr, Kinzler KW. Cancer genome landscapes. Science. 2013;339:1546–58.
71. Schumacher TN, Schreiber RD. Neoantigens in cancer immunotherapy. Science. 2015;348:69–74.
72. Segal NH, Parsons DW, Peggs KS, Velculescu V, Kinzler KW, Vogelstein B, et al. Epitope landscape in breast and colorectal cancer. Cancer Res. 2008;68:889–92.
73. Ishida Y, Agata Y, Shibahara K, Honjo T. Induced expression of PD-1, a novel member of the immunoglobulin gene superfamily, upon programmed cell death. EMBO J. 1992;11:3887–95.
74. Keir ME, Butte MJ, Freeman GJ, Sharpe AH. PD-1 and its ligands in tolerance and immunity. Annu Rev Immunol. 2008;26:677–704.
75. Okazaki T, Honjo T. PD-1 and PD-1 ligands: from discovery to clinical application. Int Immunol. 2007;19:813–24.
76. Dariavach P, Mattei MG, Golstein P, Lefranc MP. Human Ig superfamily CTLA-4 gene: chromosomal localization and identity of protein sequence between murine and human CTLA-4 cytoplasmic domains. Eur J Immunol. 1988;18:1901–5.
77. Zou W, Chen L. Inhibitory B7-family molecules in the tumour microenvironment. Nat Rev Immunol. 2008;8:467–77.

78. Snyder A, Makarov V, Merghoub T, Yuan J, Zaretsky JM, Desrichard A, et al. Genetic basis for clinical response to CTLA-4 blockade in melanoma. N Engl J Med. 2014;371:2189–99.
79. Le DT, Uram JN, Wang H, Bartlett BR, Kemberling H, Eyring AD, et al. PD-1 blockade in tumors with mismatch-repair deficiency. N Engl J Med. 2015;372:2509–20.
80. Postow MA, Callahan MK, Wolchok JD. Immune checkpoint blockade in cancer therapy. J Clin Oncol. 2015;33:1974–82.
81. Brahmer JR, Tykodi SS, Chow LQ, Hwu WJ, Topalian SL, Hwu P, et al. Safety and activity of anti-PD-L1 antibody in patients with advanced cancer. N Engl J Med. 2012;366:2455–65.
82. Hamanishi J, Mandai M, Ikeda T, Minami M, Kawaguchi A, Murayama T, et al. Safety and antitumor activity of anti-PD-1 antibody, nivolumab, in patients with platinum-resistant ovarian cancer. J Clin Oncol. 2015;33:4015–22.
83. Hamanishi J, Mandai M, Matsumura N, Abiko K, Baba T, Konishi I. PD-1/PD-L1 blockade in cancer treatment: perspectives and issues. Int J Clin Oncol. 2016;21:462–73.
84. Hamanishi J, Mandai M, Konishi I. Immune checkpoint inhibition in ovarian cancer. Int Immunol. 2016;28:339–48.
85. Naidoo J, Page DB, Li BT, Connell LC, Schindler K, Lacouture ME, et al. Toxicities of the anti-PD-1 and anti-PD-L1 immune checkpoint antibodies. Ann Oncol. 2015;26:2375–91.
86. Champiat S, Lambotte O, Barreau E, Belkhir R, Berdelou A, Carbonnel F, et al. Management of immune checkpoint blockade dysimmune toxicities: a collaborative position paper. Ann Oncol. 2016;27:559–74.
87. Tan TZ, Miow QH, Huang RY, Wong MK, Ye J, Lau JA, et al. Functional genomics identifies five distinct molecular subtypes with clinical relevance and pathways for growth control in epithelial ovarian cancer. EMBO Mol Med. 2013;5:1051–66.
88. Tothill RW, Tinker AV, George J, Brown R, Fox SB, Lade S, et al. Novel molecular subtypes of serous and endometrioid ovarian cancer linked to clinical outcome. Clin Cancer Res. 2008;14:5198–208.
89. Ketabi Z, Bartuma K, Bernstein I, Malander S, Gronberg H, Bjorck E, et al. Ovarian cancer linked to Lynch syndrome typically presents as early-onset, non-serous epithelial tumors. Gynecol Oncol. 2011;121:462–5.

MIX
Papier aus verantwortungsvollen Quellen
Paper from responsible sources
FSC® C105338

If you have any concerns about our products,
you can contact us on
ProductSafety@springernature.com

In case Publisher is established outside the EU,
the EU authorized representative is:
Springer Nature Customer Service Center GmbH
Europaplatz 3, 69115 Heidelberg, Germany

Printed by Libri Plureos GmbH
in Hamburg, Germany